22nd Surface Preparation and Cleaning Conference (SPCC 2020)

Online
31 August – 2 September 2020

ISBN: 978-1-7138-2280-6

Printed from e-media with permission by:

Curran Associates, Inc.
57 Morehouse Lane
Red Hook, NY 12571

Some format issues inherent in the e-media version may also appear in this print version.

Printed with permission by Curran Associates, Inc. (2021)

For permission requests, please contact Surface Preparation and Cleaning Conference (SPCC) at the address below.

Surface Preparation and Cleaning Conference (SPCC)
c/o Mark Thirsk
Linx Consulting-PO Box 384
Mendon, MA 01756-0384
USA

Phone: 617-273-8837

mthirsk@linx-consulting.com

Additional copies of this publication are available from:

Curran Associates, Inc.
57 Morehouse Lane
Red Hook, NY 12571 USA
Phone: 845-758-0400
Fax: 845-758-2633
Email: curran@proceedings.com
Web: www.proceedings.com

TABLE OF CONTENTS

Linx Cleaning Conferences

Monday, August 31, 2020

9:00 - 9:10 AM CT	**Introductions, Welcome, Conference Overview**	Live	**Mark Thirsk**	
9:10 - 9:45 AM	Impact of COVID-19 and outlook for Surface Prep and Cleans	Live + Q&A	Mike Corbett	Linx Consulting
9:45 - 10:20 AM	The Risks to the Consensus Economic Outlook - and What It Might Mean for Semiconductors	Live + Q&A	Duncan Meldrum	Hilltop Economics
10:20 - 10:45 AM	Wet Etch Challenges and Innovations Driving Nano-Electronics Scaling	Live + Q&A	Nabil Mistkawi	Samsung - SAS
10:45 - 11:10 AM	Wet Cleans Research and Development for Advanced Semiconductor Technologies	Live + Q&A	Brown Peethala	IBM
11:10 - 11:35 AM	Understanding the Hidden Source of Next Excursion	Live + Q&A	Dhiman Bhattacharyya	Global Foundries
11:35 AM - 12:00 PM	Future Post-CMP Cleaning Challenges	Live + Q&A	Mike White	Entegris
12:00 - 12:30 PM	Virtual Chat			
	Wrap Up / Additional Questions / Adjourn - Day 1 End		**Organizers**	

Prerecorded Presentations

	Post CMP Cleans Evolution and Defect Improvement Approaches		Katrina Mikhaylichenko	Applied Materials
	Challenges and Opportunities of Wet Clean Processes to Enable System Scaling for AI Era		Hiroaki Takahashi	Screen

Tuesday, September 1, 2020

9:00 - 9:15 AM CT	**Introductions, Day 1 Welcome, Conference Overview**	Live	**Mark Thirsk** **Joel Barnett**	
9:15 - 9:50 AM	**Keynote: Monolithic Heterogeneous 3D Integration of Germanium PMOS with Si NMOS for Enabling Moore's Law Scaling and Performance**	Live + Q&A	**Willy Rachmady**	**Intel**
9:50 - 10:10 AM	Selective Ge removal from SiGe surface for Si-cap-free SiGe passivation	Live + Q&A	Kurt Wostyn	IMEC
10:10 - 10:30 AM	Wetting of deep hydrophilic nanoholes by aqueous solutions	Live + Q&A	Guy Vereecke	IMEC
10:30 - 10:50 AM	Defect Reduction for Post Silicon Nitride Deposition Clean in MOL Sector	Live + Q&A	Bing Wu	Global Foundries
10:50 - 11:10 AM	Wet-chemical Cleaning of Cobalt and Molybdenum for Advanced Interconnects	Live + Q&A	Quoc Toan Le	IMEC
11:10 AM - 12:30 PM	**Break / Networking Reception / Poster Session**			
12:30 - 12:35 PM	**Wrap Up / Additional Questions / Adjourn - Day 2 End**		**Organizers**	

Wednesday, September 2, 2020

9:00 - 9:10 AM CT	**Day 3 Welcome**	Live	**Mark Thirsk** **Joel Barnett**	
9:10 - 9:45 AM	**Invited: CMOS Device Architecture Evolution and its Impact on Surface Preparation and Cleaning for N2 and Beyond**	Recorded	**Naoto Horiguchi**	**IMEC**
9:45 - 10:05 AM	The zeta potential of conductive material surfaces: challenges and solutions	Live + Q&A	Thomas Luxbacher	Anton Paar
10:05 - 10:25 AM	Particle analysis of unpatterned wafers with UNISERS	Live + Q&A	Altun (Ali) Ozhan	UNISERS
10:25 - 10:45 AM	Surface preparation of Niobium for Superconducting Qubit applications	Live + Q&A	Antoine Pacco	IMEC

10:45 - 11:05 AM	Advanced Purification and Defect Removal from Isopropyl Alcohol	Live + Q&A	Makonnen Payne	Pall
11:05 AM - 12:25 PM	**Break / Networking Reception / Poster Session**			
12:25 - 12:30 PM	**Wrap Up / Additional Questions / Adjourn - Day 3 End**		**Organizers**	

Prerecorded Presentations

Controlled Isotropic Etches for Gate-All-Around (GAA) Device Architectures	Yusuke Muraki	TEL
Nanosheet RMG: Metal Gate and Interface Dipole Patterning	Yusuke Oniki	IMEC
Selective TiN vs HfOx Etch with BARC Compatibility for 5 and 3nm Node Applications	Sherman Hsu	Avantor
Trimming Clean using SAPS Megasonic Technology on Xtacking 3D NAND	Sally Ann Henry	ACM Research
Wet Etching of SiO2 Film in Narrow Spaces	Dai Ueda	Screen
Study on PVA Brush Particle Loading and its Conditioning during Post CMP Cleaning	Hwi-Won Cho	Hanyang University
Mechanisms of Colloidal Ceria Contamination and Cleaning during Oxide Post-CMP Cleaning	Samrina Sahir	Hanyang University
Development of a Non-Contact Chemical Concentration Monitor	Tatsuya Nakahara	Horiba
Optimization of wet etching process to remove replacement polysilicon gate	Parth Vaidya	Global Foundries

Tuesday, September 1, 2020 Poster Session

Development of Co/Cu Wet-etch Chemistries for BEoL application	Yi-Chia Lee	EMD
Development of HA-free stripper for photoresist/ARC removal	Lili Wang	EMD
Monitoring of Fluoride in Semiconductor Cleaning Chemicals	Chuannan Bai	ECI
Isotropic Etches for Nanosheet Device Integration – Wet or Dry?	Oniki Yusuke	IMEC
Evaluating wet and dry particle deposition methods on semiconductor substrates for nanoparticle removal efficiency testing	Chimaobi Mbanaso	TEL
Challenges for chemical raw material supplies to semiconductor industry and the solutions	Hui Zhou	Huntsman
Using Wafer Edge and Bevel Vapor Phase Decomposition Inductively Coupled Plasma Mass Spectrometry for Defect Reduction in High Volume Semiconductor Manufacturing	Chris Sparks	Air Liquide
Characterization and removal of metallic contamination in process chemicals using Single Nanoparticle ICP-MS	Siddarth Sampath	Entegris
Removal of trace metals from Isopropyl Alcohol	Ashtosh Bhabhe	Entegris

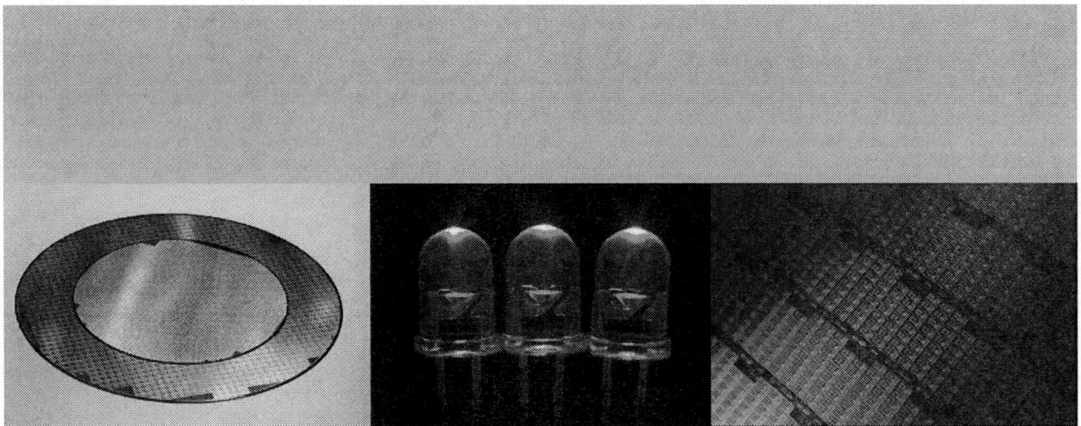

Impact of COVID 19 and Outlook for Surface Prep & Cleans

Business of Cleans 2020
August 31, 2020

Michael Corbett

Outline of Presentation

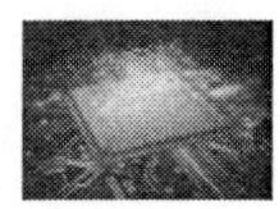

- Impact of COVID on semiconductor industry
- Trends Impacting Cleans and Surface Prep
- The Business of Cleans
- Conclusions

Impact of COVID

www.linx-consulting.com

3

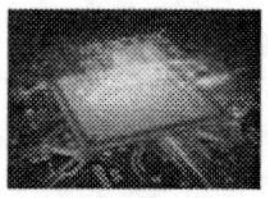

Disruptions are Increasing

The Number of supply chain shocks seems to be increasing:

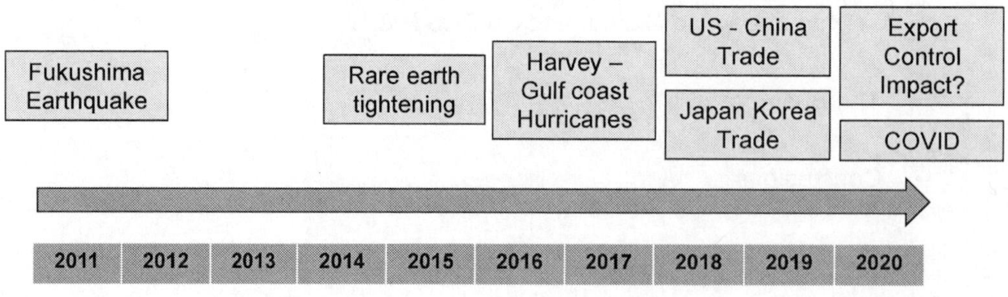

www.linx-consulting.com

4

COVID Challenges – Chemical Manufacturing

Problems	• If there is a major COVID outbreak at a non-redundant manufacturing facility and lines shut down, then there can be major supply problems for customers • Possible problematic dependence on large chemical and materials assets that are dedicated to multiple industries and not just semiconductor • Price shocks for above
Solutions	• Staggered or "A and B" shifts are increasingly being used to minimize the number of people in a plant. • This cuts into productivity and also adds cost to meet high demand or faux high demand
TBD	• The potential exists for chemical facilities to shut down due to weakness in other segments – this could have a major impact on semi in some segments such as solvents, acids and gases

www.linx-consulting.com

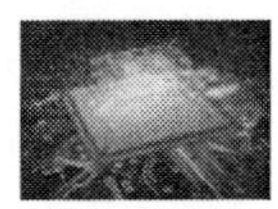

Chemical Supply - Case Studies

- Large ASUs are constructed for steel and / or petrochemical production. These units produce rare gases (Xe and Kr).
- Semi changes the Kr S/D balance

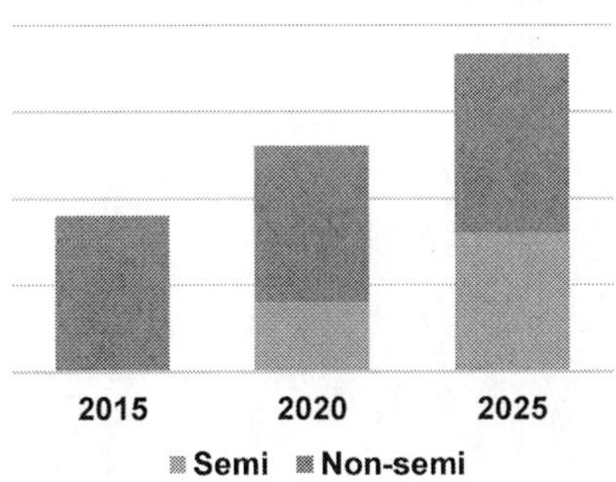

- IPA Price Shocks due to alternative demand for hand sanitizer

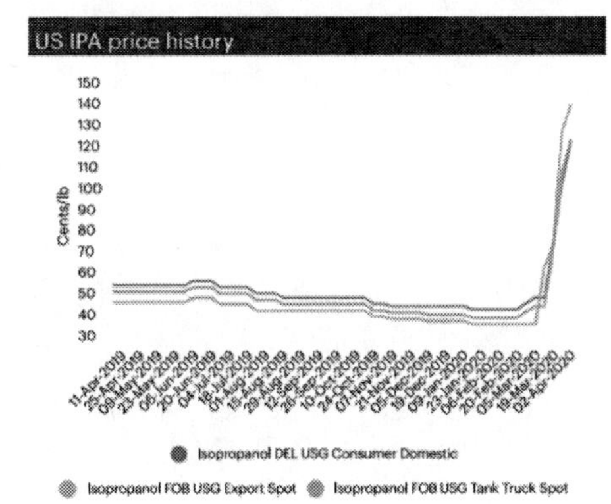

Sources: Linde and RBI ICIS

www.linx-consulting.com

Challenges - Logistics and Inventory

- Logistics were problematic at the beginning of COVID. Not all materials can be flown; shipping was a mess. The situation is getting better, but still poses challenges.
  - Need to pre-book 2 to 3 months out
  - Shipping to internal country specific locations helps larger companies as they are already holding inventory

- Problems mentioned above have caused some end-users to start building inventory for possible "COVID shocks"

- As customers are demanding excess inventory build up to deal with potential COVID shocks, stronger logistics are important to:
  - Track and return existing assets
  - Manage inventory. This is critical in order not having to buy new cylinders, etc. This also helps minimize unnecessary capex and WC requirements

- Suppliers need experts on the ground locally and by key facilities because travel is limited

Success of BCP

There remain major concerns about the robustness of upstream chemical supply for semi producers

BCP used to meet customer sustainability demands. Many MNC suppliers got a glimpse of what was happening in China first and reacted globally.

There is value gained in pressure testing the value chain to identify weaknesses. The BCP process is more robust

BCP was critical to establish redundancy. This will be more important going forward. More stability will be delivered as second source quals will be required and more multi site quals

Larger Chemcial suppliers will likely benefit from this success. A local supplier with a limited number of products or plants may struggle due to lack of redundancy and inventory costs

Major Findings BCP

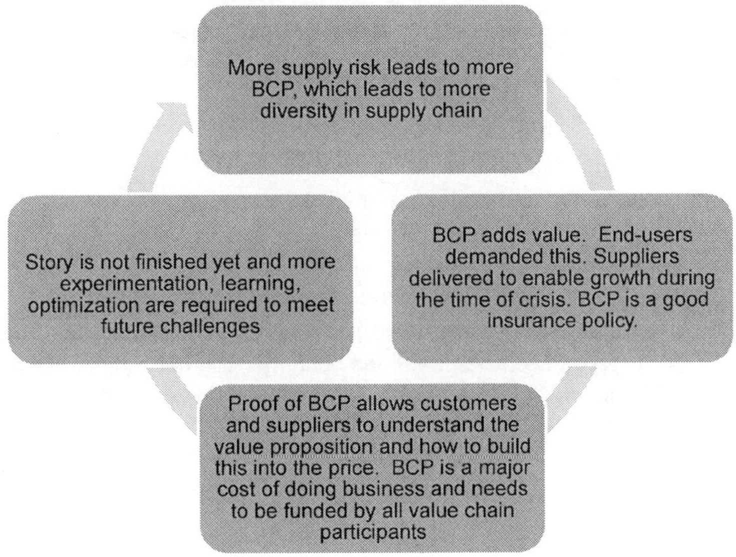

www.linx-consulting.com

9

Impact of Semi Growth During COVID

Concerns:

- End-users are worried about logistics, BCP and upstream supply shortages and supplier plant shut downs to COVID
- Top tier semi producers are showing strong growth during COVID. This growth is causing end-users to maintain strong ordering pattern for chemicals and materials
- Not all regions are providing high levels of supply and supply chain visibility
- The above reasons are making end-users start to build chemicals and materials inventory
- But are end-users ordering in excess of demand and growth outlook?

Possible Impacts:

- The side effects of inventory builds can include:
  - Growth in consignment stocks
  - Increased capital spend for cylinders, etc.
  - Increased inventory and capex costs
  - The above can have a negative impact on supplier profitability, while customers have strong profitability
  - So where do you cut costs? Analytical? Safety stocks? Investments? This can undermine the entire supply chain
- Above situation favors big suppliers and disadvantages smaller companies

www.linx-consulting.com

10

What Changes for Good?

- Key reallocation of people and assets to align with customer requirements
- Greater emphasis on BCP and its proven value
- R&D and Development
  - With continued growth and scaling, customers need more development activities and more R&D engagement
    - Pre COVID it was more FTF
  - Now expect more virtual collaborations. This benefits customer.
    - For example - previously the Taiwan team engaged with TSMC.
    - In a virtual environment, the customer can get access to a broader range of assets to maximize benefits and you can have more of a global team.
    - Broader and better collaborations with global bench involved. However stronger firewalls will also be required

www.linx-consulting.com

11

Trends Impacting Cleans & Surface Prep

www.linx-consulting.com

12

New FEOL Materials Challenge Cleans

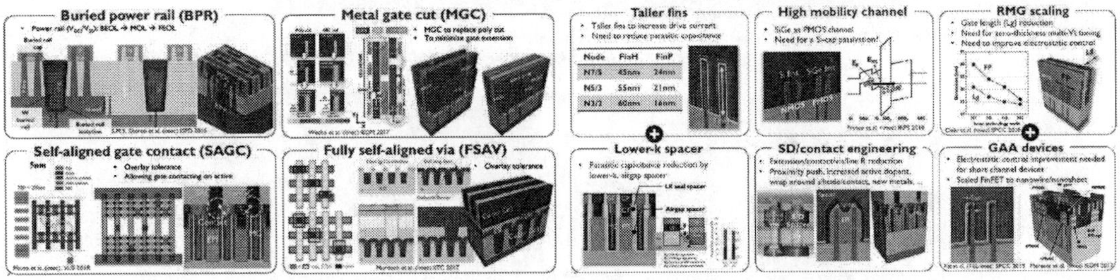

- Taller fins
- Construction of buried power rails
- Metal gate cut (MGC) processing to drive gate density
- Self-aligned gate contacts
- Fully self-aligned via
- High-mobility channel materials
- Scaling replacement metal gate materials and process
- Low-κ spacers
- Improved source/drain contacts
- Development of gate-all-around (GAA) architectures

Source: SPCC 2019/IMEC; Coventor

www.linx-consulting.com

13

High Selectivity Nitride Strip

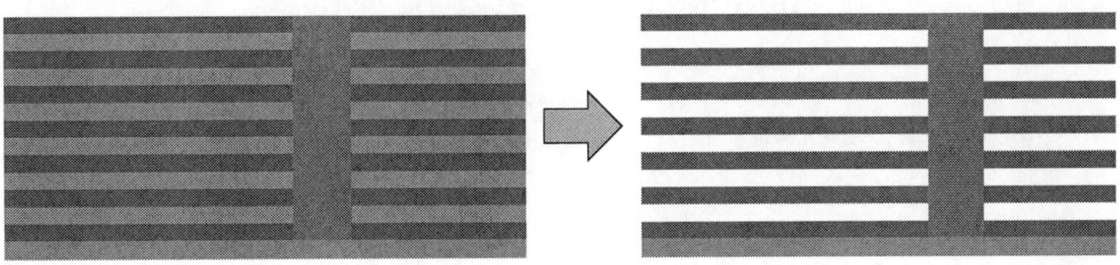

- HSN is still done in batch process tool for 3DNand
- This process be long duration > 2 hours to be complete
- Advanced chemistries are required to decrease long residency times

www.linx-consulting.com

14

High Aspect Ratio Drying Challenges

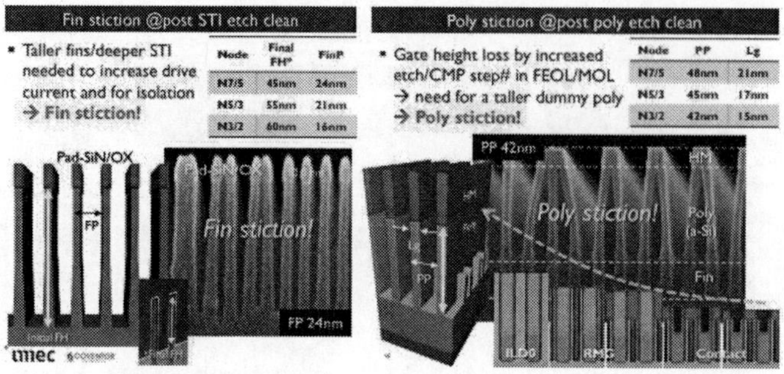

- Stiction and pattern collapse issues are getting worse with successive nodes and HAR structures
- Use of IPA may be extendable; though increasing issues with purity
- $SCCO_2$ has been introduced in DRAM
- Sophisticated drying schemes involving several solvents

Source: SPCC 2019/IMEC/Coventor

www.linx-consulting.com

15

Filtration Remains Critical

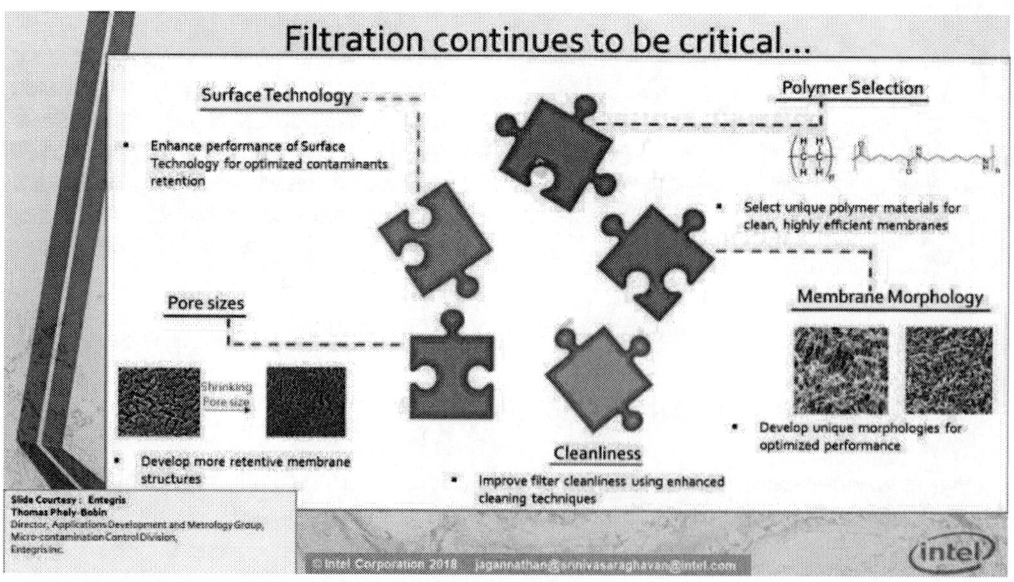

Source: Intel Business of Cleans, 2018

www.linx-consulting.com

16

Post-CMP Requirements & Challenges

Requirements:
- Removal of particles, debris and organic residues
- Removal of metal ions
- Minimal corrosion and etching
- No dendrite growth
- Mitigate copper oxide growth on
- No modification of the dielectric
- No increase in line resistivity
- No impact on reliability
- Safety and ESH
- Long shelf life and pot life
- Low cost of ownership

Challenges:
- New materials/film types
  - Diversification of dielectric materials
  - New barrier materials
- Feature sizes decreasing
- Increasingly complex CMP slurries
  - New particle types
  - Small particles
  - Advanced organic additives
- Defect detection thresholds decreasing

Sources: Entegris and DuPont EKC

RMG CMP in FinFET Structure

Reduction in defectivity can be achieved by:
- Morphology and purity of slurry abrasives
- Reduction in slurry abrasive content in slurry formulation
- Advanced polishing pads to reduce micro-scratches and other defects
- Reduction in down force for polishing to have minimum abrasion
- Use of ultra-high purity chemicals for slurry formulation
- Role of CMP cleaner module in reducing defects

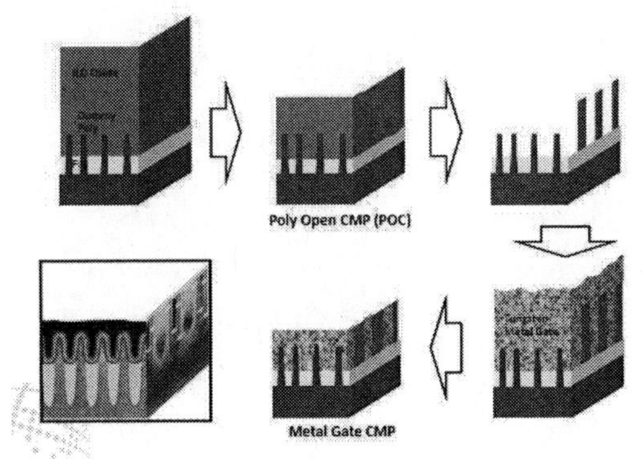

Source: Hong Jin Kim, GF,CAMP 2016

Ceria Slurry Development

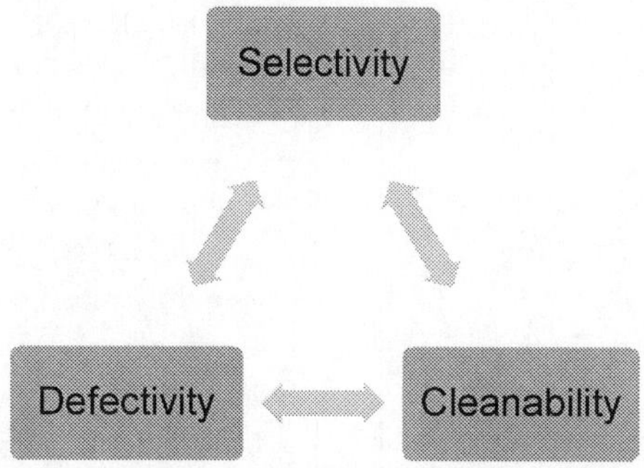

- FEOL CMP demands atomic level CMP control with minimum material removal with high selectivity. Removal rate becomes less critical for advanced processes because a desired approach is to use lower downforce to obtain better control of dishing and erosion.
- Colloidal ceria has been considered critical for advanced node CMP, primarily for defectivity and better process control.
- A major issue with colloidal ceria remains cleanability.

www.linx-consulting.com

19

The Business of Cleans

www.linx-consulting.com

20

Surface Prep Chemistry and Systems ~ $7B

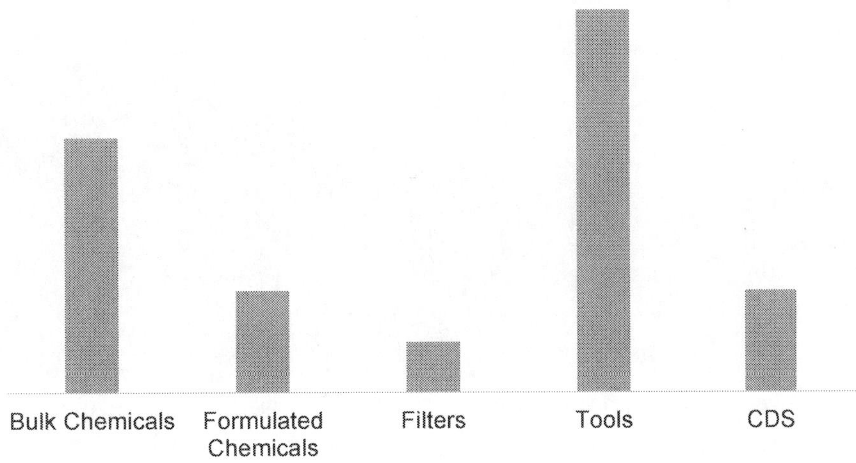

www.linx-consulting.com

Materials Intensity Increases

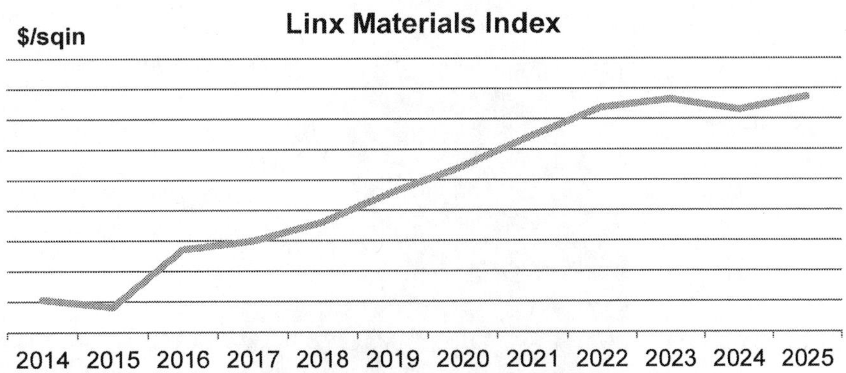

Advanced devices require high BOM costs and drive up materials intensity as more wafers move to advanced designs.

www.linx-consulting.com

Specialty Formulated Cleaning Chemicals

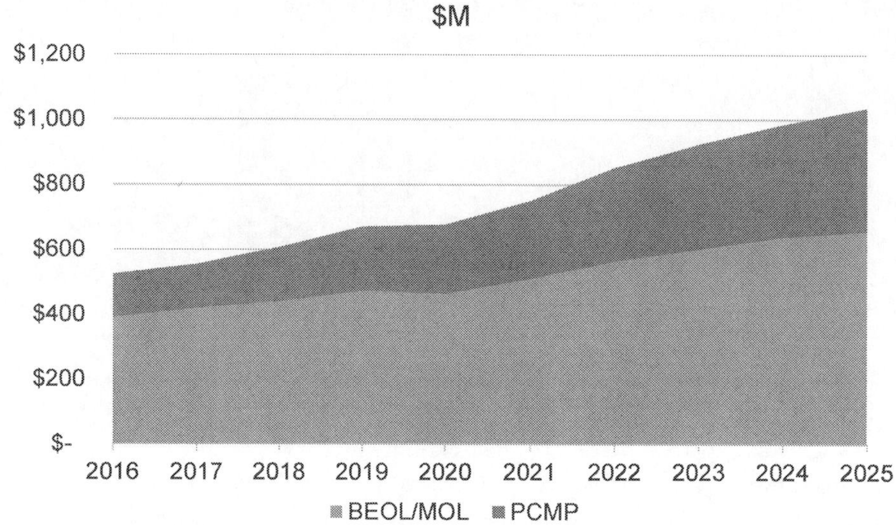

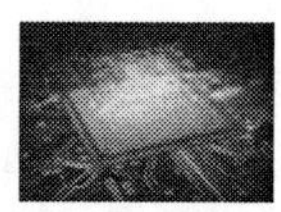

Leading Suppliers Formulated Cleans

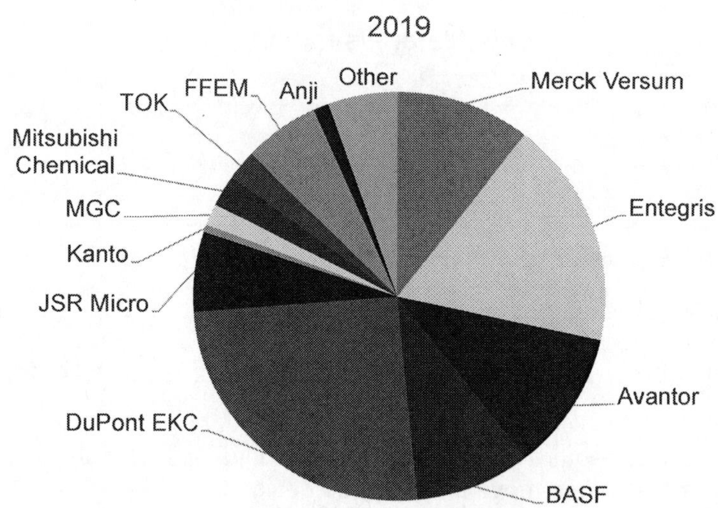

Materials Suppliers - Future Directions

- Advanced node / leading edge manufacturing has the highest capacity utilization and is driving industry growth during the COVID Pandemic
- Advanced node integration is becoming increasing more customer specific based on each semiconductor producer's unique integration
- 10nm is seen as an inflection point for many materials suppliers
  - Quality
  - Supply chain
  - Purity
  - Metrology
  - New capabilities becoming more critical

www.linx-consulting.com

25

Conclusions

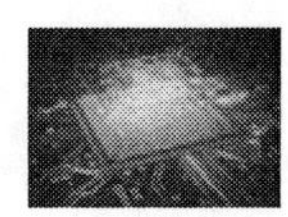

- BCP becomes more critical as the number of shocks is increasing
- This recession does not look like past recessions. It will impact different suppliers in different ways based on their respective portfolios.
- Materials proliferation raise concerns of increased number of chemicals and compatibility challenges. Product development becomes increasingly more customer specific
- Similar to fab and OEM consolidation, the materials suppliers will continue to consolidate and COVID will push this process

www.linx-consulting.com

26

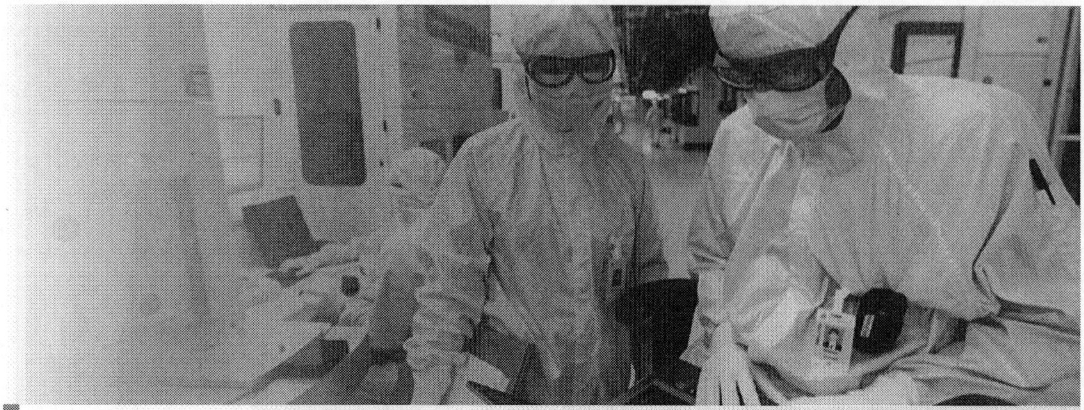

Understanding the hidden source of the next excursion

Dhiman Bhattacharyya

GLOBALFOUNDRIES
Malta, NY - 12020

August 31, 2020

GLOBALFOUNDRIES®

The Business of Cleans Conference, 2020

Outline

- **Transformation of the IC industry**
- **Cost of contamination**
- **Sources of contamination**
- **Contamination classification**
- **Contamination mitigation**
- **Summary**
- **Acknowledgement**

Transformation of the IC industry

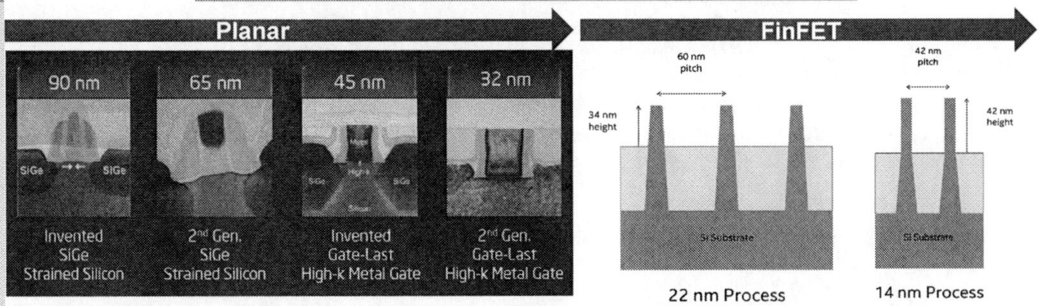

	Year of Production	2013	2015	2017	2019	2021
1	Logic Industry "Node Name" Label	16/14	10	7	5	3.5
2	Logic ½ Pitch (nm)	40	32	25	20	16
3	FinFET Fin Half-pitch (new) (nm)	30	24	19	15	12
4	FinFET Fin Width (new) (nm)	7.6	7.2	6.8	6.4	6.1
5	FinFET Fin Width (nm) (*Detailed*)	6.4	5.3	4.4	3.7	3.1

ITRS Road Map 2013

Transformation of the IC industry

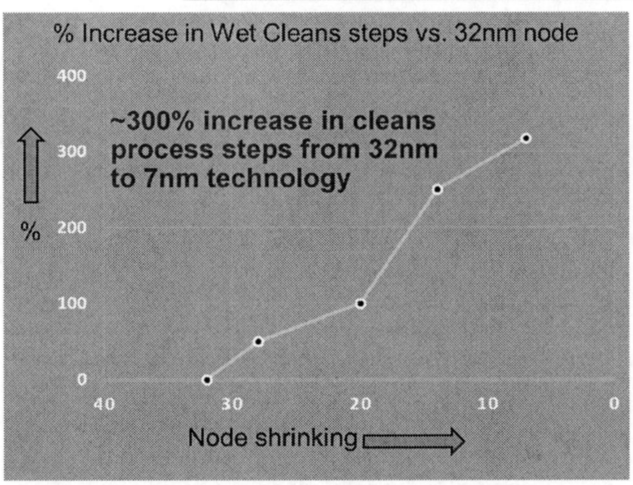

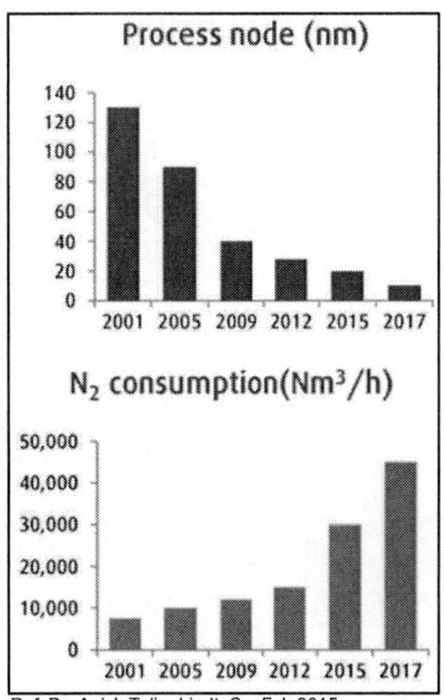

- Enormous increase in the number of process steps as the technology is shrinking over the years.
- Consequently, consumption of bulk chemicals is also sky rocketing in parallel.

Ref: Dr. Anish Tolia, Lindt, ConFab 2015

Translation of Particle Defects

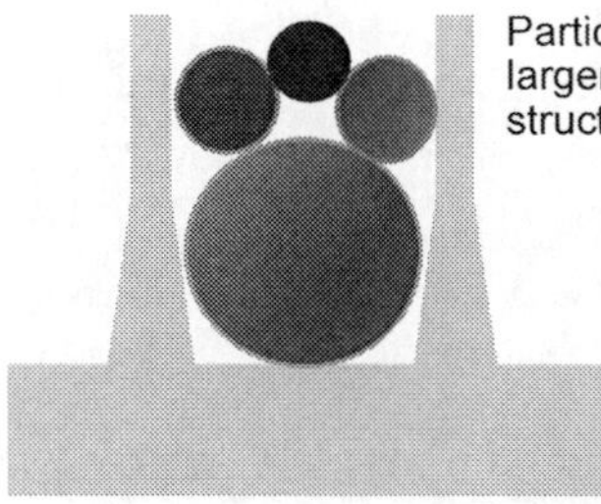

Particles on larger pitch structures

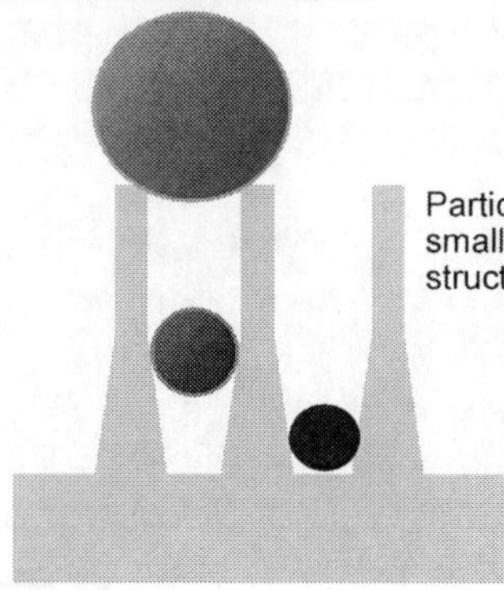

Particles on smaller pitch structures

Particles on vias/trenches

The Business of Cleans Conference, 2020

Cost of contamination

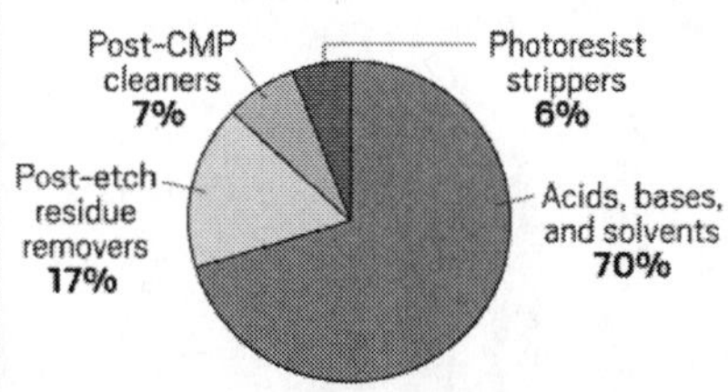

2017 market = $2.13 billion

Chemical & Engineering News
Vol 95, Issue 40 | pp. 32-33, 2017

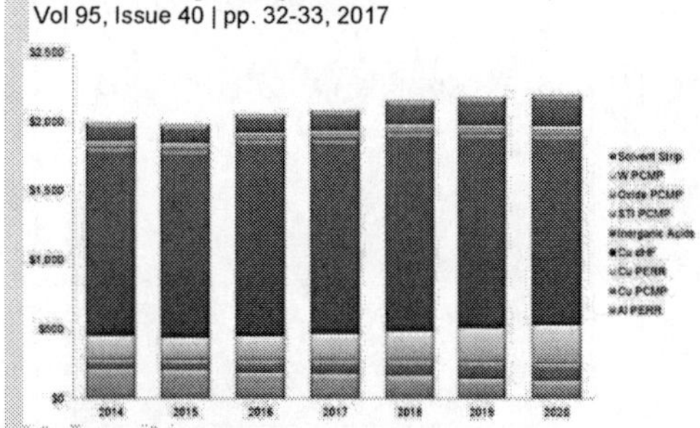

Bulk Wet Chemical Report, LINX Consulting

- Acids, bases and solvents constitutes >70% of the chemical supply
- Sulfuric acid is mostly used acid
- Hydrogen peroxide is highest in volume
- An ISO = ~ 16,000L of peroxide
- Peroxide = $1-$2/L
- *1ISO = $16K - $32K*

Hypothetical calculation:

- Average usage = 500ml/wafer
- Wafers exposed = 8000
- 1 Wafer = ~$1000 - $10,000
- *$ Impact from 1 ISO = $8M - $80M*

The Business of Cleans Conference, 2020

Cost of contamination

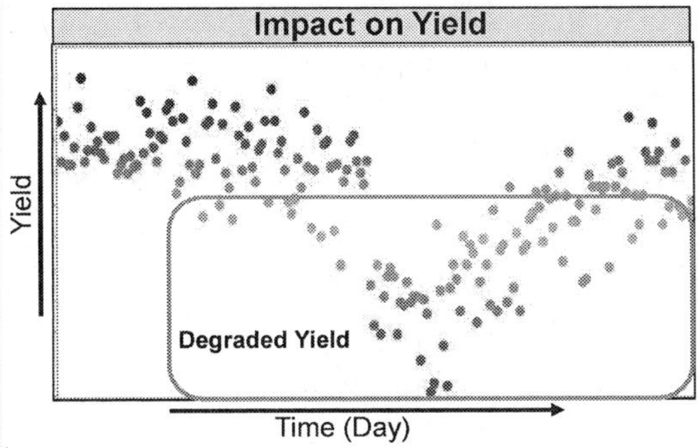

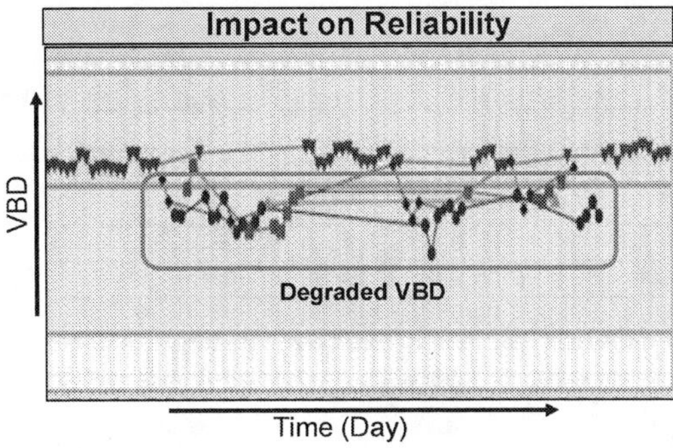

- A sudden burst of sub-standard quality of materials directly reflects in SORT Yield and reliability.
- Above plots show Yield vs. Process Date (Left Graph), and VBD vs. Process Date.
- A surge of incoming material with a very high quantity of sub 30nm particles remained undetected at 30nm and impacted the wafers yield.
- A surge of incoming material with a very high quantity of organics remained undetected and impacted the wafer's reliability (VBD).
- It takes days to recover the production and return to baseline.

The Business of Cleans Conference, 2020

Cost of contamination

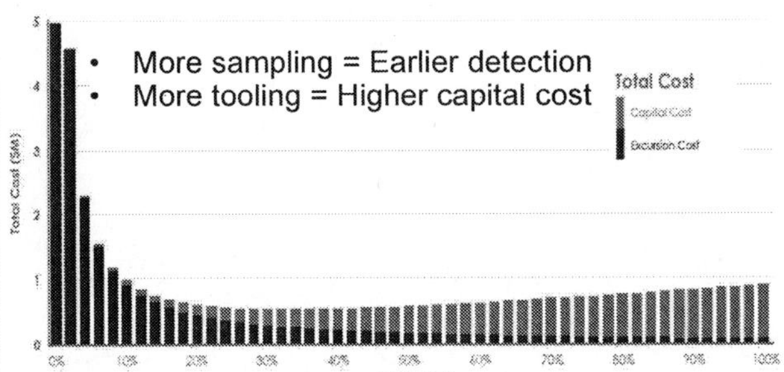

D. W. Price and D. G. Sutherland in Process Watch: Sampling matters

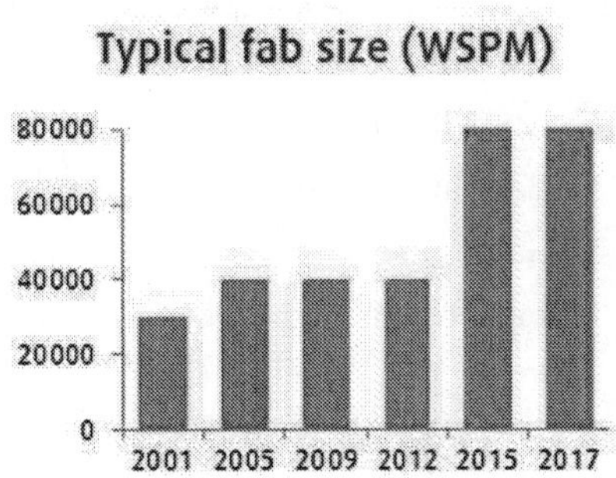

Dr. Anish Tolia, Lindt, ConFab 2015

- Number of wafer starts is increasing every year
- Any excursion becomes disastrous for HVM, customer chains get impacted as well
- Additional engineering time for troubleshooting, tool down times and additional chemical for line flushing significantly add to the manufacturing cost.
- Larger sampling rate helps in earlier detection, however additional tooling involves huge capital investment and it also slows down the wafer movement.

The Business of Cleans Conference, 2020

Sources of contamination

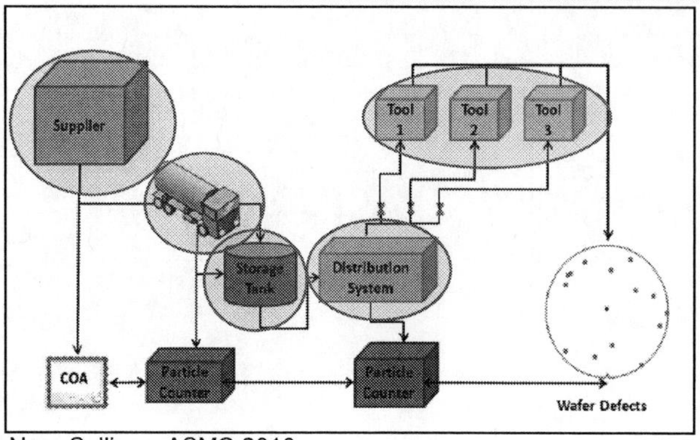

Nora Colligan, ASMC 2018

Initial suspects (on-site):

- Storage units
  - Supplying containers
  - In-house storage tanks
- Supply lines
- Pumps/engines
- Valves
- T-Boxes
- Filters

Secondary suspects (supplier's site):

- Storage units
- Supply lines
- Filters
- Manufacturing column/mixing tanks
- Raw materials
- Packaging

Contamination classification

5 major classes of contaminants

1. Particles
2. Metallic ions
3. Chemicals
4. Bacteria
5. Airborne molecular contaminants (AMCs)

The effects of contamination on semiconductor devices

1. **Device processing yield** - contaminants may change the dimensions device parts ⇒ reduce overall yield.
2. **Device performance** – Latent defects may potentially alter device dimensions or material quality - high amount of mobile ionic contaminants in the wafer can change the electrical performance of the device.
3. **Device reliability** - Failure of device due to the presence of a small amount of metallic/ particulate contaminants => travel inside the device and end up in electrically sensitive areas and cause device failure.

Contamination classification

Allowable 20nm particles/ml in chemicals (*ITRS*)

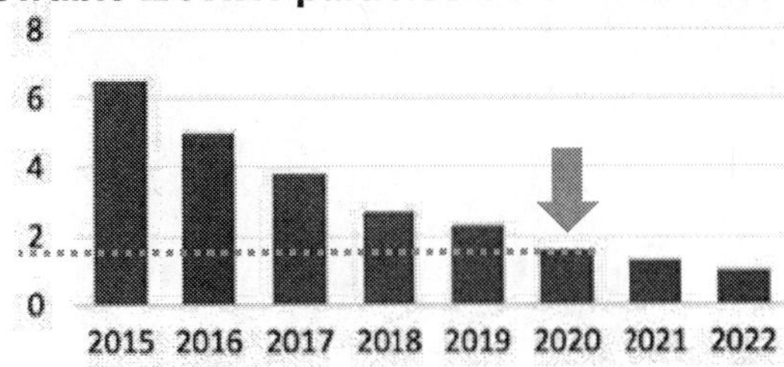

Are we there yet with the required bulk chemical quality?

Challenges:

- **Metrology**
  - 20nm liquid particle monitoring is available in the market. How many implemented?
  - Awaiting next generation metrology of on-the-wafer detection
- **Cost of ownership**
  - \$/Transistor vs. \$/Gal

Contamination classification

Particles

Quality

- Chemical composition
- Physical shape & size
- Size distribution

Quantity

- Number of particles in each of the size distribution

Quality:

- Chemical composition varies depending on the nature of the failure
  - PTFE/PFA filter breakdown => C & F
  - Pump O-ring breakdown => Si & C
  - Manufacturing process => SiOx
- Physical shape/size depends on the filtration capability
- Size distribution also depends on filtration capability

Contamination classification

Quality:

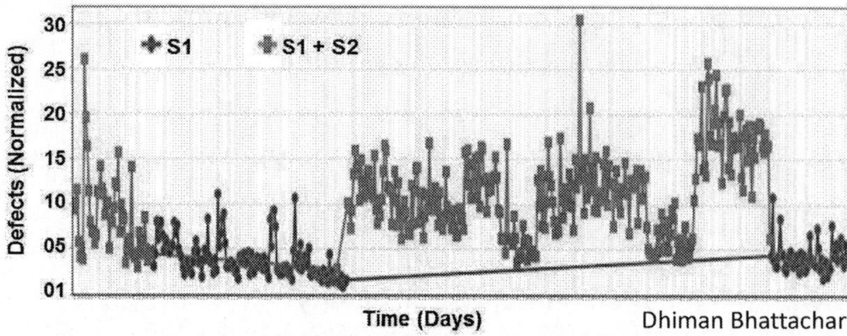

Dhiman Bhattacharyya, SPCC 2017

Example of impact from a difference in chemical manufacturing processes:

- Sulfuric acid supply in bulk from both the suppliers (S1 and S2).
- The defect baseline came down when supply from S2 was interrupted for few months.
- Baseline trended up again when supply from S2 was restored.
- Defect level again trended down after supply from S2 was stopped.
- S2 was found to have glass lined column while S1 had Teflon lined column.
- Small SiOx particles from S2 supply created the defects on the product wafers.

The Business of Cleans Conference, 2020 GLOBALFOUNDRIES 13

Contamination classification

Quantity:

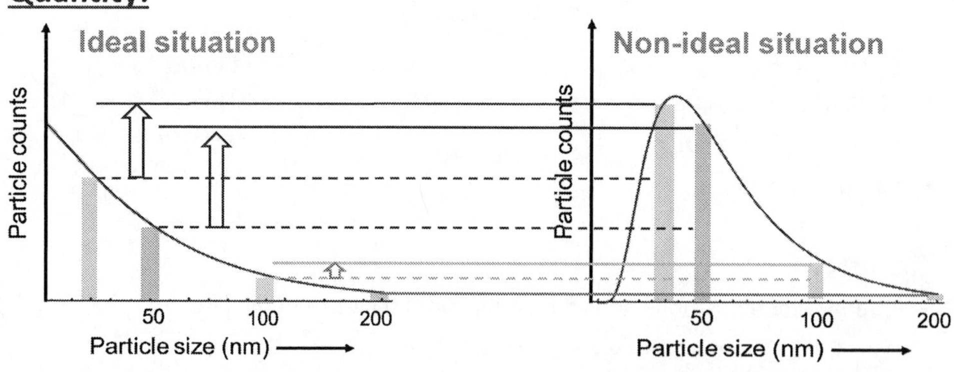

Conventional understanding leads us to a conclusion where we assume that a stable reading @ 100nm counts indicates a consistent quantity of sub-100nm particles as well.

In reality, the correlation between the particle counts between different sizes may not follow normal distribution.

The Business of Cleans Conference, 2020 GLOBALFOUNDRIES 14

Contamination classification

Quantity:

Variation from packaging

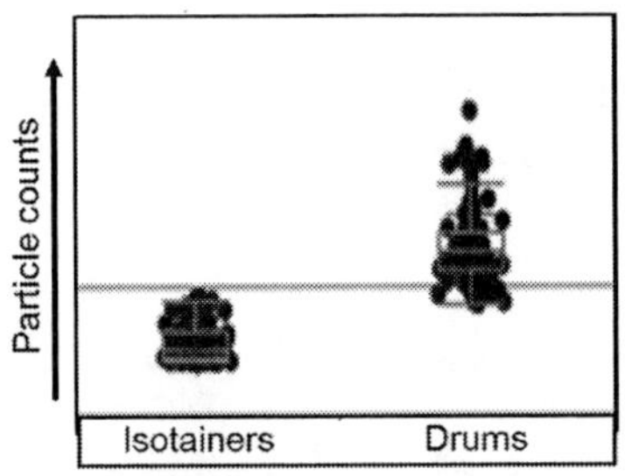

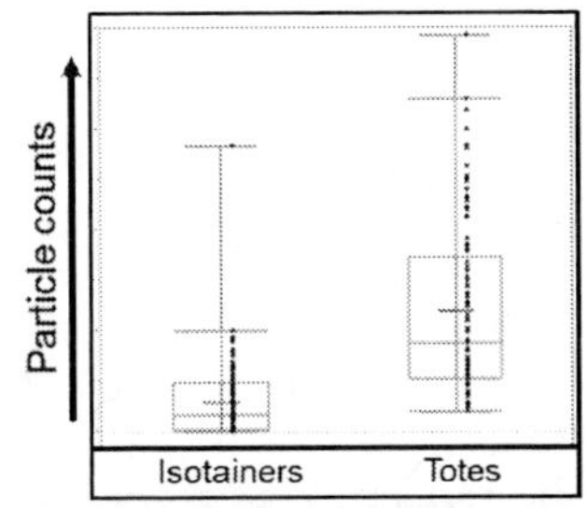

Containers	Usual Sampling Rate	Ideal Sampling Rate	Reason
Isotainers	1 data point/grab sample post fill	Continuous LPC monitoring through ISO during fill	To build confidence on the integrity of the material
Totes/ Drums	1st and last tote/drum of the batch	Sample from each unit of tote/drum fill	To avoid occasional tote/drum failures

The Business of Cleans Conference, 2020

GLOBALFOUNDRIES 15

Contamination classification

Quantity:

Variation from suppliers

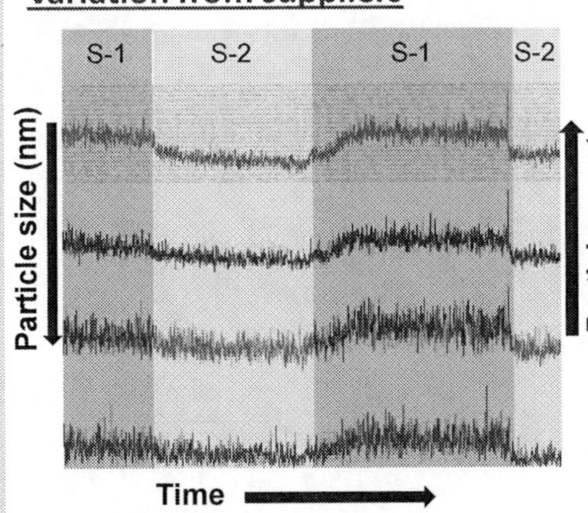

- An example of the ideal variation in particle size distribution from supplier to supplier.
- Particle counts of all particle sizes moved up or down by change in suppliers.

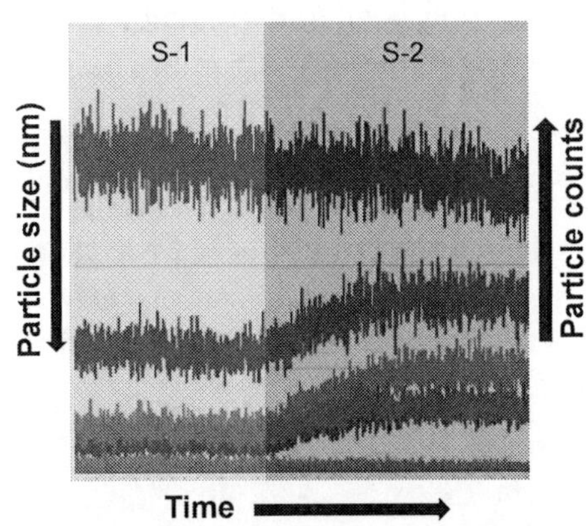

- An example of the non-ideal variation in particle size distribution from supplier to supplier.
- Particle counts of some particle sizes moved up while smallest particles' count trended up by a change in suppliers.

The Business of Cleans Conference, 2020

GLOBALFOUNDRIES 16

Contamination mitigation

The biggest challenge in bulk chemical supply is inconsistency in quality:

- *Too many sources of contaminations*
  - Raw materials, UPW
  - clean room
  - packaging containers and filling process
- *Opportunities to improve quality control*
  - Process control management: Real time Process Control systems
  - Quality check & SPC control of raw materials, sub-suppliers
  - Advanced metrology detection & test methods validation
  - Contamination free manufacturing: only PTFE/PFA wetted surfaces & BIC filtration
  - Holistic SPC program: not limited to COA
- *Lack of dedication to semiconductor grade supply*
  - Opportunities to stick to semicon customers, not relaxing the spec to sell to non-semicon customers
- *Lack of standards across the suppliers*
  - Mfg. process, storage and packaging differences
- *Lack of standards across the customers*
  - Differences in technology offerings, subsequent requirements and associated cost
- *Volatility in demands based on consumer markets*
  - Inconsistent chemical demands due to instability in semicon market

Summary

- Ever moving industry is running at its own pace to beat or meet Moore's law.
- Bulk chemical related excursions are becoming too frequent and too expensive to HVM and following customer chain
- Most of the bulk chemicals already meet the trace metal ions spec. However, they are far behind their particle spec.
- Particle monitoring should be looked at the distribution of each possible size ranges, not as a big bucket of particles because distribution may vary from one size to other and they may have different impacts.
- More sensitive and faster technologies need to be evaluated to detect and analyze these smaller defects.

Acknowledgement

To all my colleagues whom I work with and work for at GLOBALFOUNDRIES.

I would like to specially thank the following people:

- Thayalan Kulasingam
- Kevin Prettyman
- Vincent Sih

Dhiman Bhattacharyya
GLOBALFOUNDRIES
400 Stonebreak Rd. Ext., Malta, NY - 12020
Mobile: 518-948-3123/ 817-676-5633
Email: dhiman.bhattacharyya@globalfoundries.com

Defect Reduction for Post Silicon Nitride Deposition Clean in MOL Sector

Bing Wu
Senior Process Engineer
GLOBALFOUNDRIES
9/1, 2020

2020 SPCC Virtual Conference

1

Presentation Outline

- **Introduction to MOL sector**
- Problem Statement
- Results and Discussion
- Conclusions
- Future Work

2

Function of the MOL

The Mid Of Line provides:

- **Strain layers (optional) for enhanced device performance**

- **Interlayer dielectric (ILD) insulation between devices and BEOL**

- ***W contacts to***
  → gates
  → source / drain regions
  → resistors
  → substrate diodes

  (resistors and substrate diodes: nonFETs)

3

Simplified MOL Process Flow

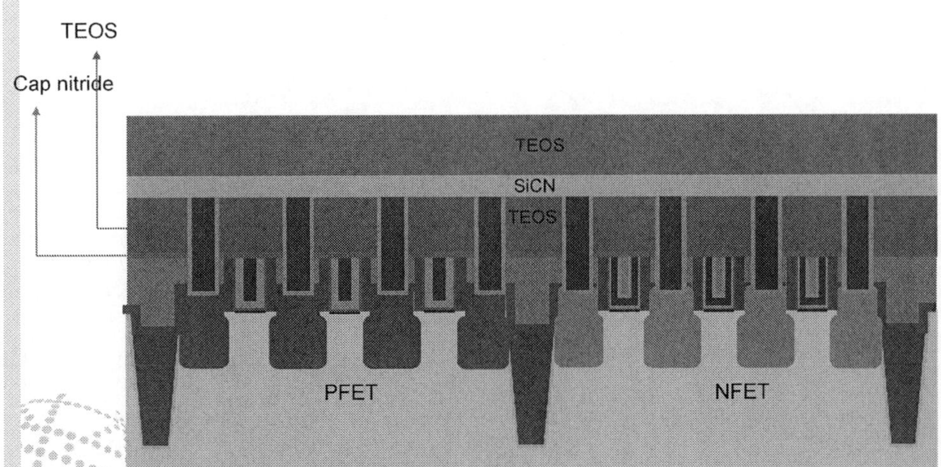

Simplified MOL Process Flow

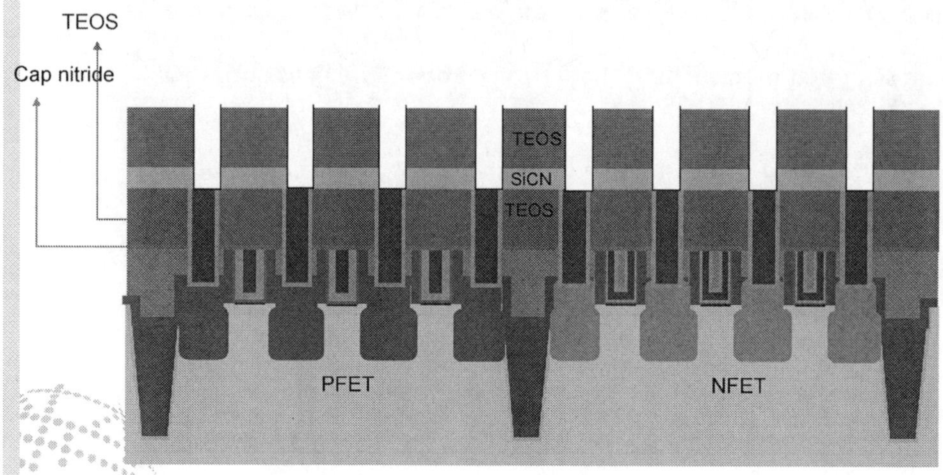

Simplified MOL Process Flow

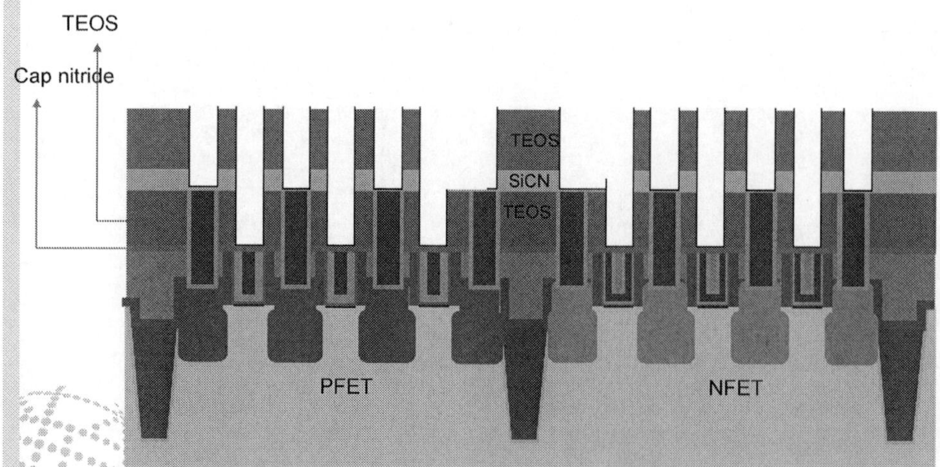

Simplified MOL Process Flow

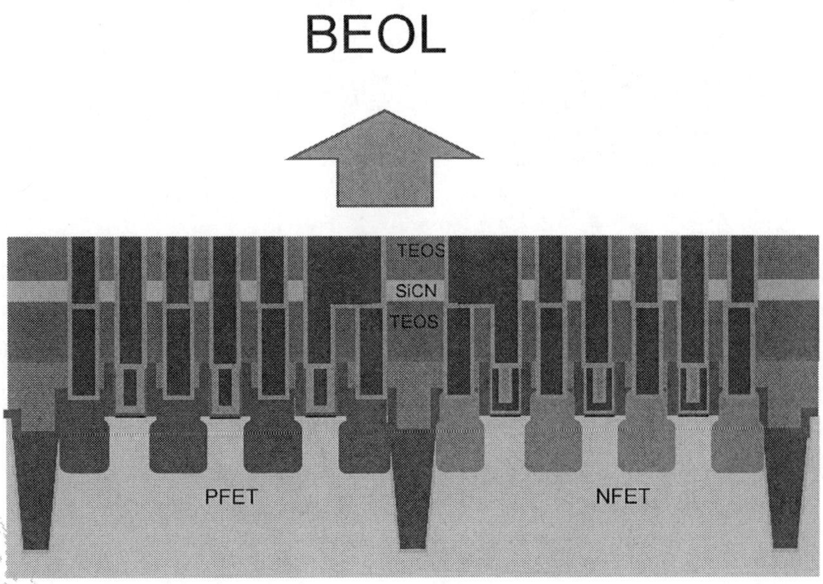

Presentation Outline

- Introduction to MOL sector
- **Problem Statement**
- Results and Discussion
- Conclusions
- Future Work

8

Problem Statement

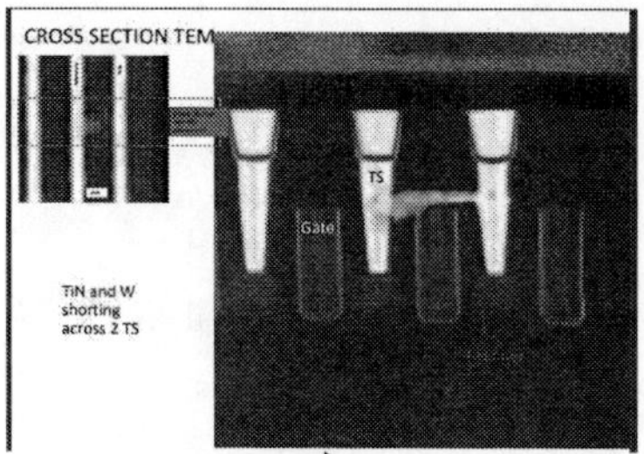

TS-PC short

❖ Proposed mechanism
Defects on nitride surface etched away doing patterning, generating run paths for liner/metal movement

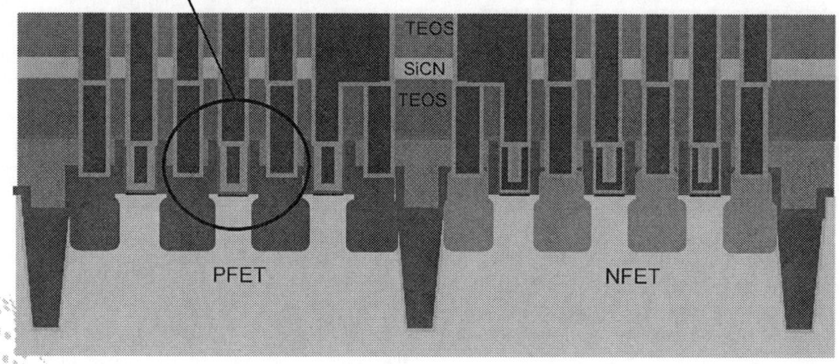

Defects generation during nitride deposition

Surface particles

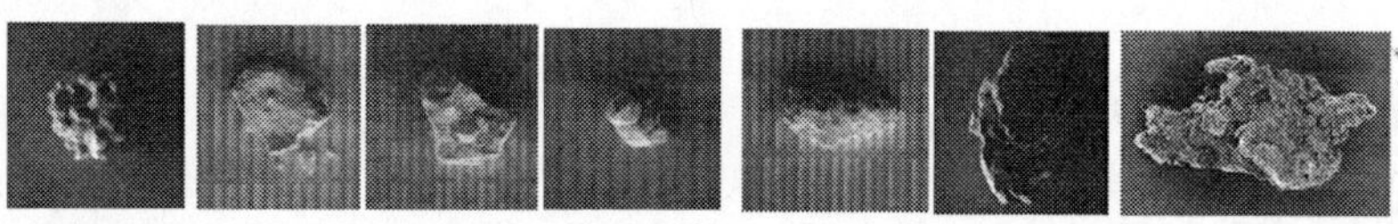

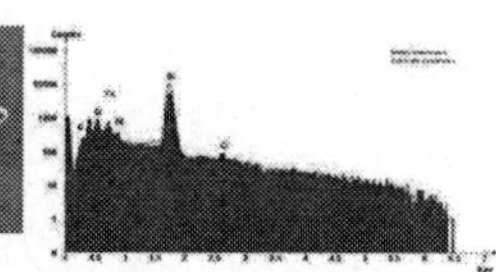

*not to scale

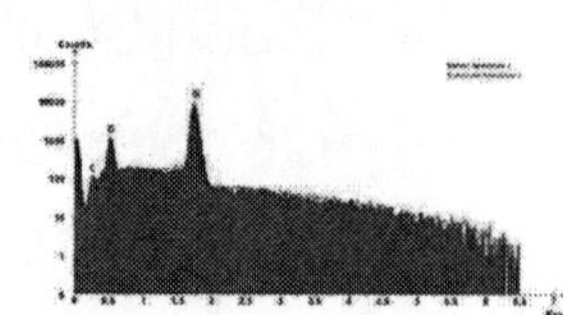

Organic particles

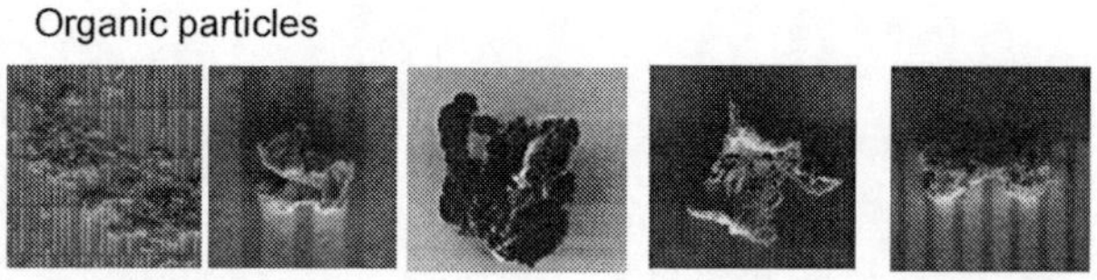

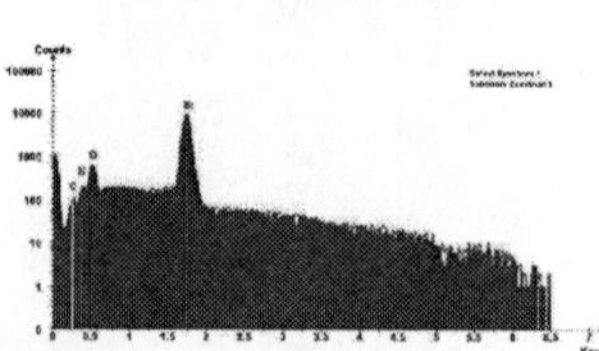

*not to scale

- ❖ Surface particle and organics are non exclusive, they often classified together
- ❖ Both defects sometimes contain minor amount of metals, such as Al, Fe, Ni

10

Post nitride deposition clean important

- Post nitride deposition requires a wet clean process to remove surface defects

Proposed Solution

- Improve post nitride deposition cleaning by focusing on organic related defects removal
- POR chemistry: basic chemistry with oxidizer
  pH 8-11, good for particle removal
- Goto chemistry: acidic chemistry with oxidizer
  pH 2-4, good for organic removal

11

Presentation Outline

- Introduction to MOL sector
- Problem Statement
- **Results and Discussion**
- Conclusions
- Future Work

12

Surface Particle Reduction

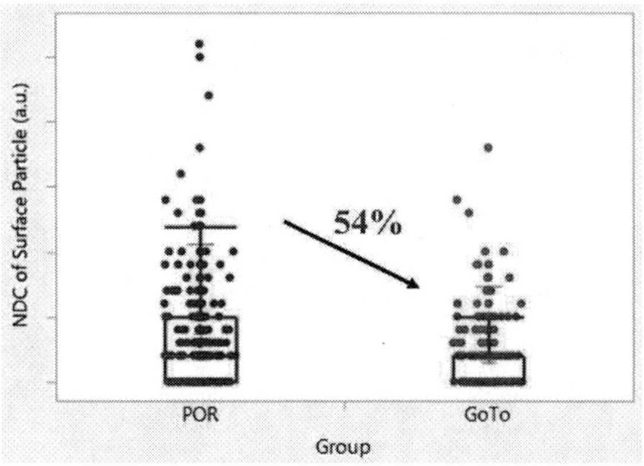

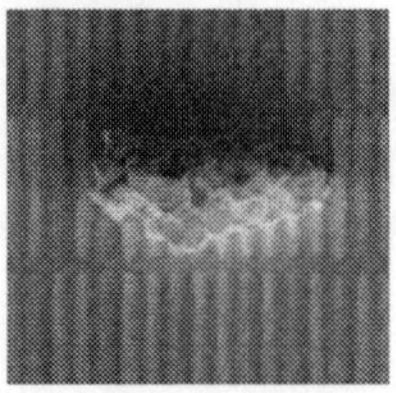

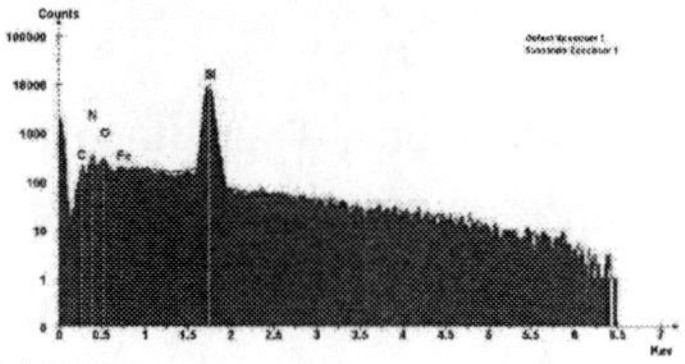

- ❖ Goto chemistry significantly reduces surface particles
- ❖ Silicon nitride surface is hydrophobic in nature, attracting particles
- ❖ Goto chemistry effectively oxidize nitride surface to form a thin oxide layer, reducing hydrophobicity

13

Organic Reduction

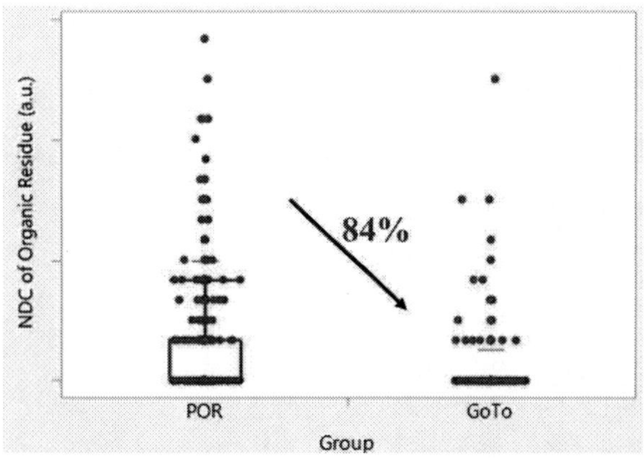

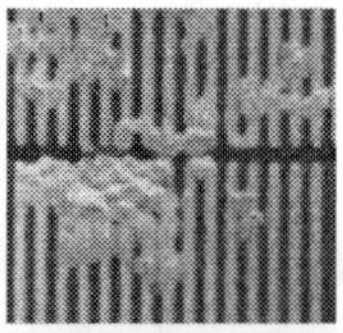

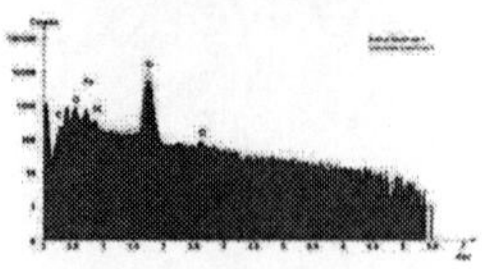

- ❖ Defects contains significant organic component evident from the carbon, nitrogen and oxygen peaks and some metals
- ❖ Goto chemistry able to oxide organic component, forming soluble –OH/-COOH surface groups, enhance particle removal

14

Embedded Defects Reduction

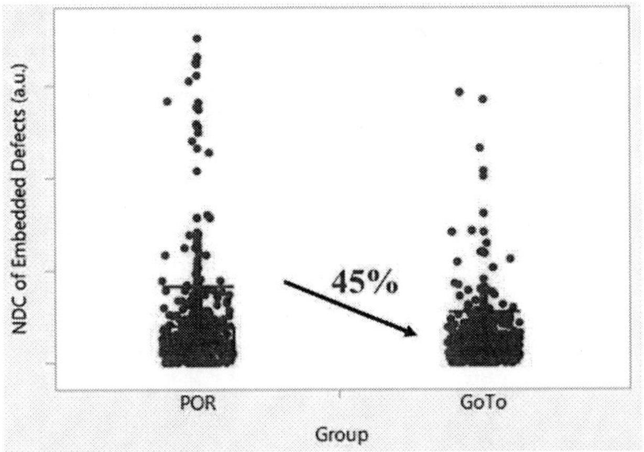

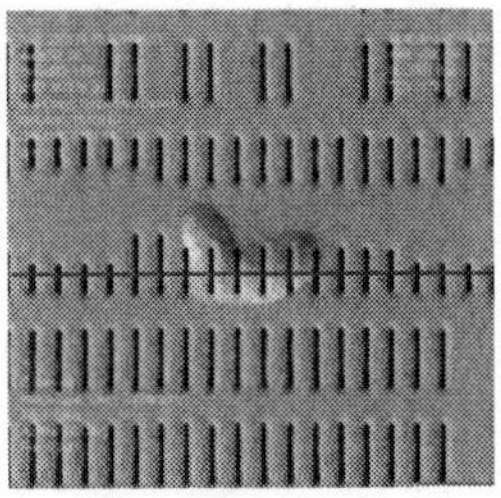

- Embedded defects coming from surface defects during nitride deposition, coated with subsequent layer deposition
- TEOS deposition reveals smaller defects not picked up in prior scans

15

Yield Improvement

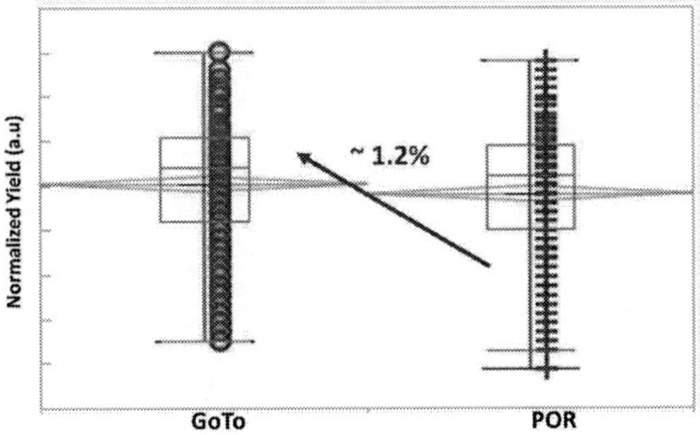

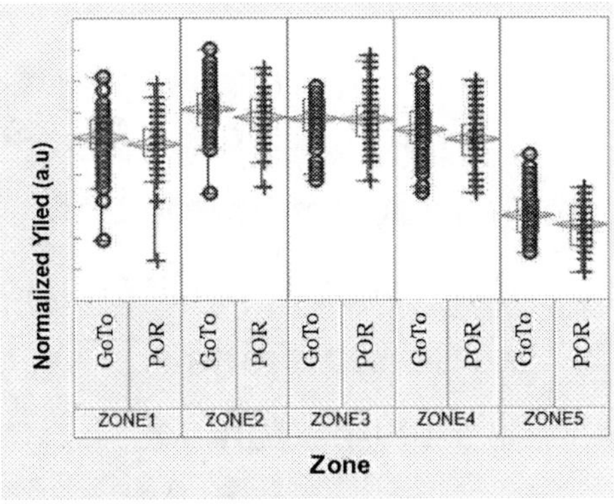

- Defect reduction translate to yield improvement
- Goto chemistry increase yield by ~1.2%
- Yield improvement consistent across all zones from wafer center to edge

16

TS-PC shorts

- Shorts statistically comparable to POR, no significant improvement

17

Presentation Outline

- Introduction to MOL sector
- Problem Statement
- Results and Discussion
- **Conclusions**
- Future Work

18

Conclusions

- Goto chemistry reduces surface particles, organics significantly
- Surface defect reduction leads to less embedded defects
- Defect reduction translates to ~1.2% yield enhancement
- Yield improvement consistent from wafer center to edge
- TS-PC short not improved based on current data

19

Future Work

- Process recipe improvement (such as dispense time, flow rate, spin speed)

20

Thank you

21

Hilltop Economics

Extreme Uncertainty and the Economic Outlook* Upstream Optimism VS Downstream Pessimism

Business of Cleans – SPCC 2020

August 31, 2020

Duncan Meldrum
Hilltop Economics LLC
duncan.meldrum@hilltopeconomics.com

*Based on Consensus Economics, Inc. CONSENSUS FORECASTS®

Hilltop Economics

Outline

- Pandemic shock and economic uncertainty
- Final Demand (real GDP) consensus* outlook
- Monthly data tracking
  - To find the bottom of the economic decline
  - To reconcile upstream optimism with downstream pessimism
- Scenarios for planning in uncertain times
  - Aggressive tech, consensus
  - Normal behavior, consensus
  - Extended Pandemic

*Consensus: I calculate a consensus global real GDP forecast using world bank history and consensus forecasts for 85 countries reported by Consensus Economics® in their four regional monthly Consensus Forecasts. Consensus Economics surveys more than 700 forecasters monthly to produce their Consensus Forecasts. www.consensuseconomics.com

Hilltop Economics

The Covid-19 Pandemic Produces a Severe Shock

Almost 25 million cases still growing about 1%/day; more than 825,000 deaths (>3%)

New Cases By Day Through August 28

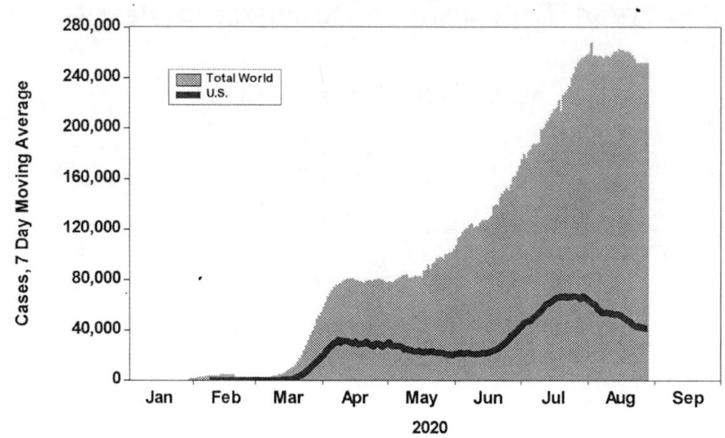

Hilltop Economics

Pandemic Control Produces Uncertainty at a High Level

	Timing	Effectiveness	Economic Impact
Social Distance (Masks, personal hygiene precautions, etc.)	Varies	Varies	Varies by country
Curtail Economic Activity	Varies	Varies	Severe
Develop Effective Treatment Protocol	Months since inception	Varies, still relatively limited	Minor
Develop Vaccine	End 2020 at the earliest	Unknown	Dependent on Effectiveness

Hilltop Economics

Responses to the Pandemic Disrupt Economic Policy

Global Economic Policy Uncertainty

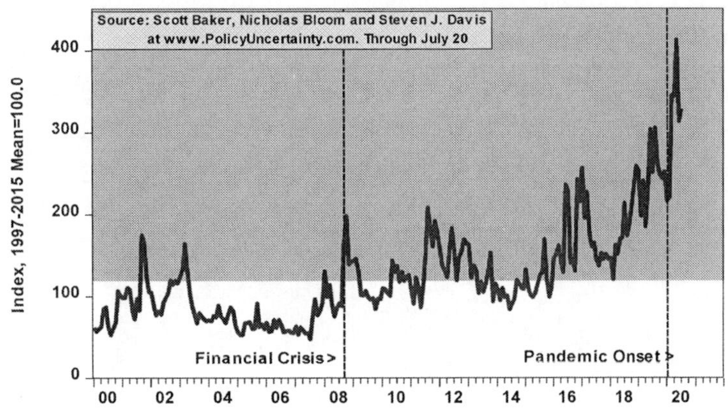

5

Hilltop Economics

Dealing With Uncertainty

- Normal forecasting techniques break down in severe shock events
- Pandemic and economic policy response introduce multiple uncertainty levels
- Planning & operating outlooks:
  - Consensus forecasts lag events for a while, but tend to be the best over the long run
  - Real-time tracking of higher frequency data
  - Scenarios to assess the future

6

Hilltop Economics

Consensus Outlook

The Consensus "V" Outlook for Final Demand

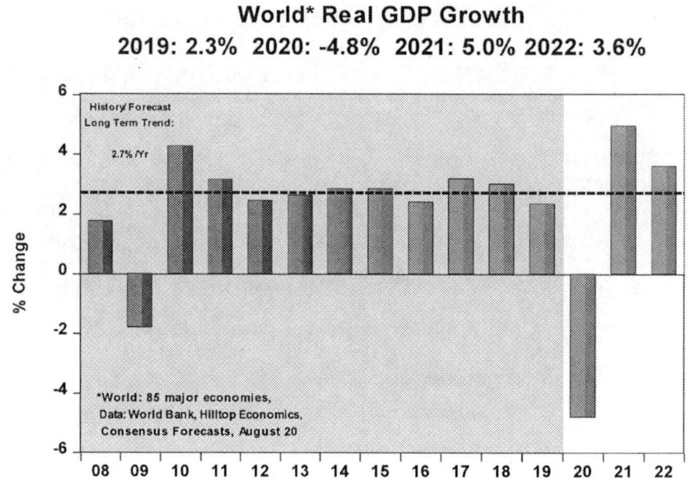

Hilltop Economics

Consensus Outlook

Unusually Wide Swing in Consumer Spending
Investment Cycle More Typical, But Much Weaker Recovery Expected

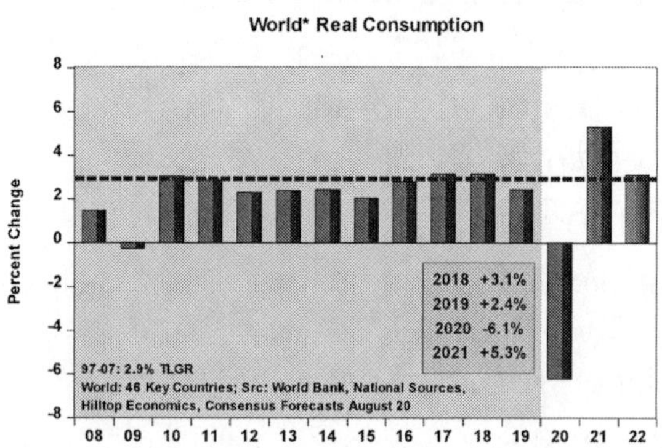

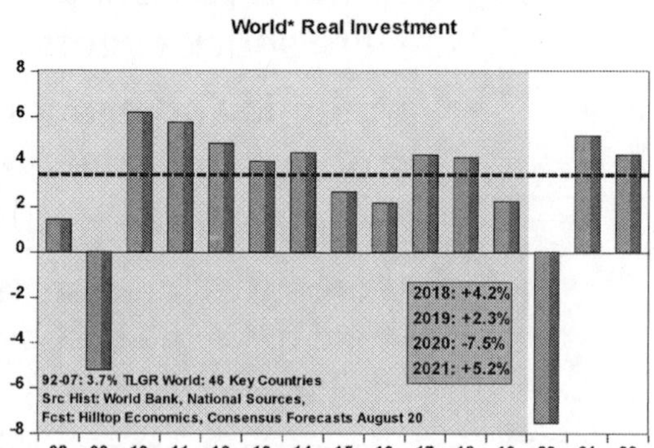

Hilltop Economics

Near the trough?

The Evolution of the Consensus Outlook Since January

Consensus-Based World Real GDP Forecasts
August Forecast: 2020: -4.8% 2021: +5.0%

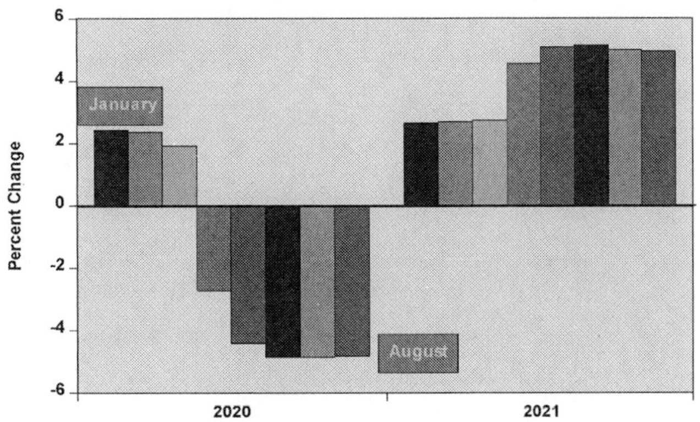

Forecast By Month Made: January 20 to August 20

9

Hilltop Economics

Near the trough?

Narrowing Range of Uncertainty in 2020 Real GDP Outlook

Real GDP Consensus Forecast® (August)	2020 Percent Change	Lowest to Highest	2020 RANGE August	2020 RANGE July	2020 RANGE June	2019 RANGE in June 2019
U.S.	-5.2 ^	-6.8^ to -3.3	3.5	3.9	6.2	0.5
Eurozone	-7.9 ^	-11.0^ to -6.3v	4.7	5.9	6.9	0.7
Japan	-5.3 v	-6.6^ to -3.1	3.5	3.5	4.0	0.6
China	2.1 ^	0.5^ to 3.3	2.8	3.8	5.8	0.6
So Korea	-1.1 v	-2.0^ to 0.3	2.3	2.4	3.8	1.1
Taiwan	0.0 ^	-2.6 to 1.8^	4.4	4.3	6.7	0.8
Brazil	-6.2 ^	-9.0 to -4.2	4.8	4.8	4.9	1.3
Russia	-4.4 ^	-6.6^ to -1.5^	4.1	6.0	6.1	1.1

^ Forecast increased; v Forecast decreased
Source: Consensus Forecasts® Surveys of International Forecasts, August, July, June 2020 & June 2019

10

Hilltop Economics

Beyond the trough?

A Deep Recession, But LEI Rose in May through July

11

Hilltop Economics

Beyond the trough?

U.S. Unemployment Peaked in April, Rising Elsewhere - But Less of a Problem

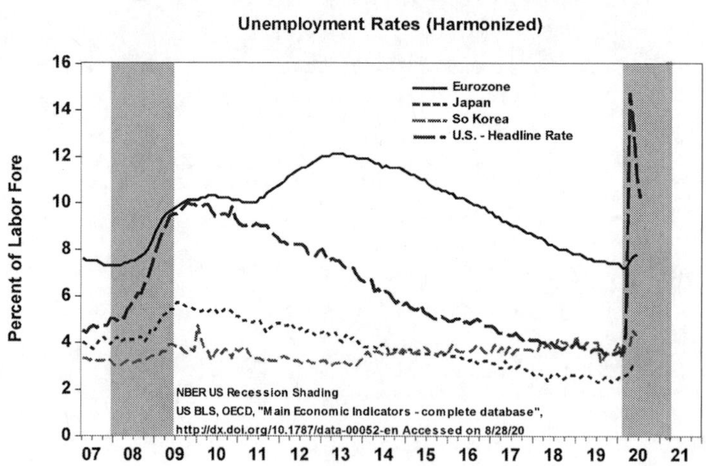

12

Weak Confidence, Improving Except in China & US

Consumer Confidence Indexes

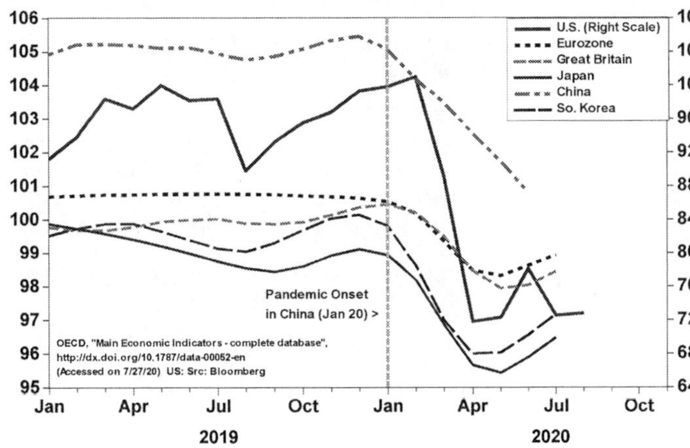

Industrial Production & Manufacturing

Monthly Performance Indicators	AVE %CHYA	Percent Change VS Year Ago											
Production & Shipments data	Prior 5 Yrs	2019M08	2019M09	2019M10	2019M11	2019M12	2020M01	2020M02	2020M03	2020M04	2020M05	2020M06	2020M07
China Industrial Production (Value Added Basis	6.4%	4.4%	5.8%	4.7%	6.2%	6.9%	(13.5%)	(13.5%)	(1.1%)	3.9%	4.4%	4.8%	4.8%
Japan Industrial Production Index	0.4%	(3.0%)	(1.1%)	(6.9%)	(6.6%)	(6.5%)	(2.4%)	(3.7%)	(6.8%)	(15.9%)	(24.5%)	(21.0%)	
Korea Industrial Production Korea (SA)	1.0%	(2.7%)	1.4%	(1.0%)	1.4%	6.5%	4.4%	3.0%	6.5%	(2.2%)	(8.3%)	(2.0%)	
U.S. Industrial Production Index	1.2%	0.3%	(0.2%)	(0.8%)	(0.4%)	(0.8%)	(0.8%)	(0.2%)	(4.6%)	(16.3%)	(15.8%)	(11.0%)	(8.2%)
Germany Industrial Production Index	1.1%	(3.9%)	(4.7%)	(5.7%)	(3.5%)	(6.0%)	(3.1%)	(2.6%)	(14.0%)	(30.3%)	(23.8%)		
IP Manufacturing - Japan	0.5%	(3.0%)	(1.2%)	(6.8%)	(6.6%)	(6.6%)	(2.3%)	(3.5%)	(6.9%)	(16.0%)	(24.5%)	(20.9%)	
IP Manufacturing - Korea	1.0%	(2.5%)	1.3%	(0.9%)	1.5%	6.9%	4.8%	3.3%	6.9%	(2.3%)	(8.5%)	(2.1%)	
IP Manufacturing - Taiwan	2.8%	2.4%	(0.5%)	(3.0%)	2.1%	6.7%	(1.9%)	21.5%	11.9%	4.9%	1.8%	7.6%	2.6%
IP Manufacturing - Singapore	3.2%	(6.7%)	(1.1%)	2.9%	(11.2%)	(1.7%)	12.1%	(9.5%)	20.9%	11.4%	(5.0%)	(8.8%)	(8.9%)
IP Manufacturing U.S. (NAICS)	1.0%	(0.3%)	(1.0%)	(1.5%)	(0.7%)	(1.1%)	(0.7%)	(0.2%)	(5.1%)	(19.7%)	(16.6%)	(10.9%)	(7.5%)
IP Manufacturing - Euro Area	1.8%	(2.3%)	(1.7%)	(2.1%)	(1.2%)	(2.9%)	(1.6%)	(2.0%)	(14.2%)	(30.4%)	(21.8%)		
IP Manufacturing - Germany	1.0%	(3.8%)	(4.8%)	(5.7%)	(3.5%)	(6.0%)	(3.2%)	(2.7%)	(13.9%)	(30.1%)	(23.6%)		
I-S Japan Manufacturing	(2.5%)	(6.0%)	(3.4%)	(7.7%)	(10.0%)	(7.6%)	(9.2%)	(6.5%)	(14.1%)	(31.2%)	(38.9%)	(25.9%)	
I-S Korea Manufacturing	(2.9%)	(4.5%)	(5.3%)	(6.0%)	(2.5%)	7.8%	(4.8%)	(6.2%)	(3.5%)	(7.5%)	(14.8%)	(6.4%)	
I-S Ratio Taiwan Manufacturing	(2.3%)	(8.7%)	(5.4%)	0.0%	(0.2%)	6.3%	(11.8%)	14.2%	3.8%	(5.9%)	(12.7%)	(0.1%)	
I-S U.S. Manufacturing	(1.0%)	(3.4%)	(3.9%)	(4.1%)	(3.5%)	(2.7%)	(2.5%)	(2.8%)	(7.3%)	(23.1%)	(19.7%)	(9.6%)	
Number of Rising indicators		3	3	3	4	6	3	4	5	3	2	2	2
Number of Falling indicators		13	13	13	12	10	13	12	11	13	14	11	3
Net (Rising minus Falling)		(10)	(10)	(10)	(8)	(4)	(10)	(8)	(6)	(10)	(12)	(9)	(1)

Source: National Statistical Agencies
I-S Ratios: Growth Rates are INVERSE (i.e. Shipments/Inventories)- negative rate of change indicates RISING I-S ratio

Hilltop Economics

Monthly Tracking System

Intermediate Demand: Products With Semiconductors

Monthly Performance Products Industrial Production (Volume index) or units	AVE %CHYA Prior 5 Yrs	2019M07	2019M08	2019M09	2019M10	2019M11	2019M12	2020M01	2020M02	2020M03	2020M04	2020M05	2020M06	2020M07
		Percent Change VS Year Ago												
Computers:														
Korea IP (SA) Mfg of Computers and Peripheral Equipment	(3.5%)	(12.9%)	(25.2%)	1.5%	(16.0%)	(16.9%)	(20.5%)	(13.8%)	(15.3%)	(16.0%)	(0.8%)	(14.6%)	(4.6%)	
Taiwan IP Mfg: Computers	25.9%	131.6%	109.6%	84.0%	45.0%	87.0%	40.7%	9.7%	10.2%	25.3%	24.4%	50.0%	77.6%	
Taiwan ODMs Revenue (Compal, Quanta, Inventec)	#DIV/0!	(4.5%)	(10.4%)	0.5%	(2.3%)	(5.6%)	(7.2%)	(23.1%)	(25.9%)	(1.1%)	14.0%	6.0%	7.5%	21.5%
Taiwan Portable computer (#)	52.3%	71.7%	105.0%	119.6%	118.1%	244.2%	262.1%	15.3%	27.8%	26.9%	91.6%	261.0%	223.6%	
Taiwan Industrial computer (#)	10.0%	10.7%	(7.0%)	(1.6%)	6.1%	(0.2%)	27.2%	(8.1%)	29.6%	6.7%	(7.2%)	(6.4%)	17.4%	
Taiwan Server (#)	9.1%	81.2%	74.2%	84.6%	107.1%	70.5%	36.0%	22.3%	63.6%	73.7%	79.5%	50.0%	115.4%	
China PC Output (Million)	(1.2%)	(0.2%)	(6.6%)	7.2%	7.2%	4.0%	17.0%	(31.4%)	(31.4%)	3.5%	25.8%	28.4%	4.4%	20.0%
Japan IP 6204 Computer Production	(3.4%)	34.3%	60.5%	72.2%	54.4%	38.6%	42.3%	31.2%	20.1%	(7.9%)	19.4%	(14.3%)	(20.1%)	
Japan Notebook Units	3.0%	48.3%	77.7%	114.5%	60.8%	42.1%	37.8%	25.3%	2.9%	(8.9%)	16.0%	6.5%		
Japan Desktop Units	(3.0%)	51.1%	72.6%	126.5%	67.9%	78.9%	75.9%	102.6%	31.4%	2.8%	21.8%	(8.1%)		
Japan PC Server Units	15.2%	37.3%	7.0%	27.0%	10.8%	3.9%	3.6%	2.9%	(32.6%)	(1.3%)	(42.3%)	(44.5%)		
U.S. IP (SA) Computers and peripheral equipment 3341	6.3%	(2.8%)	(0.0%)	(0.5%)	(2.4%)	1.0%	1.0%	0.5%	0.6%	0.4%	(1.7%)	(6.6%)	(4.0%)	(1.9%)
Peripherals:														
Taiwan IP Mfg: Other Computer Peripheral Equipment	9.6%	24.2%	5.0%	(8.3%)	2.6%	6.0%	20.5%	(17.8%)	27.0%	12.9%	16.2%	20.1%	24.1%	
Taiwan IP Mfg: Monitors and Terminals	4.7%	(7.8%)	(18.8%)	10.1%	(15.3%)	(14.1%)	(13.8%)	(22.8%)	77.9%	(26.9%)	(21.3%)	(13.5%)	(2.3%)	
SGP IP Computer Peripherals & Data Storage (33412)	(2.7%)	(1.3%)	9.7%	4.6%	2.2%	(14.3%)	(2.8%)	1.3%	(30.2%)	(22.6%)	(24.7%)	(17.5%)	(38.8%)	(21.1%)
Mobile Phone/Communications:														
Korea IP (SA) Mfg of Telecommunication and Broadcasting Apparatuses	(12.7%)	(22.6%)	20.9%	7.2%	10.5%	8.2%	(1.9%)	5.4%	0.9%	(1.9%)	(24.7%)	(26.6%)	(24.1%)	
China Mobile Phone Output (Million)	2.9%	3.9%	(2.3%)	7.6%	4.2%	3.9%	4.1%	(41.2%)	(41.3%)	(18.8%)	(16.0%)	(20.5%)	(14.1%)	(12.3%)
Taiwan IP Mfg: Telephones and Cellular Phones	(34.9%)	(51.7%)	(42.5%)	19.5%	(13.9%)	(29.6%)	(38.5%)	(30.9%)	33.1%	(31.0%)	27.1%	11.9%	(14.7%)	
Taiwan IP Mfg: Other Communication Equipment	6.7%	42.0%	59.2%	51.4%	29.7%	25.8%	30.7%	(1.4%)	29.3%	13.9%	8.1%	1.6%	19.4%	
U.S. IP (SA)Communications equipment 3342	2.5%	12.3%	9.7%	8.8%	8.7%	9.1%	6.6%	1.3%	(2.1%)	(0.5%)	3.5%	7.3%	6.9%	6.7%
Video/Audio Equipment:														
Korea IP (SA) Mfg of Electronic Video and Audio Equipment	(17.5%)	(8.3%)	(9.3%)	(12.9%)	(12.6%)	(12.6%)	2.6%	(7.4%)	(15.5%)	(3.7%)	(11.3%)	(28.1%)	(5.5%)	
Taiwan IP Mfg: Audio and Video Equipment	3.7%	7.6%	(0.8%)	0.2%	14.4%	16.6%	21.6%	(3.2%)	67.2%	(5.0%)	25.7%	(3.9%)	15.5%	
U.S. IP (SA)Audio and video equipment 3343	3.6%	37.0%	32.7%	45.1%	39.4%	28.5%	6.3%	7.0%	3.3%	(2.4%)	(8.1%)	(21.2%)	(13.6%)	(9.0%)
Navigation/Measuring/Electromedical/Etc.														
U.S. IP (SA)Navigational, measuring, electromedical, and control instrum	1.8%	3.2%	3.9%	6.3%	5.4%	6.2%	4.9%	5.8%	8.6%	5.5%	(1.9%)	(3.7%)	1.1%	1.6%
Taiwan IP Mfg: Measuring, Navigating and Control Equipment	6.9%	20.1%	29.0%	21.4%	(5.4%)	8.5%	3.3%	(11.7%)	16.4%	16.0%	3.0%	(4.1%)	(14.9%)	
Taiwan IP Mfg: Irradiation and Electromedical Equipment	9.4%	9.8%	10.4%	(17.6%)	34.9%	(12.2%)	21.7%	(21.6%)	9.5%	34.0%	37.5%	41.6%	71.6%	
Consumer (Other and Miscellaneous):														
SGP IP Infocomms & Consumer Electronics	(4.0%)	3.5%	3.0%	7.4%	(1.0%)	3.8%	(14.3%)	33.7%	(19.0%)	(12.4%)	(2.3%)	5.5%	0.5%	(16.6%)
Japan IP 6203 Consumer electronics 61.4	(1.7%)	(7.2%)	(11.2%)	(9.7%)	(16.0%)	(20.4%)	(25.5%)	(16.2%)	(21.4%)	(30.7%)	(31.7%)	(37.8%)	(52.3%)	
Number of Rising Indicators		18	17	22	19	19	20	14	18	12	15	12	13	4
Number of Falling Indicators		10	11	6	9	9	8	14	10	16	13	16	12	5
Net (Rising minus Falling)		8	6	16	10	10	12	0	8	(4)	2	(4)	1	(1)

Hilltop Economics

Monthly Tracking System

Semiconductor Production

Strong growth everywhere except Japan (weak) and Europe (declining)

Monthly Performance Semiconductors Industrial Production/Volume except WSTS (Revenue)	AVE %CHYA Prior 5 Yrs	2019M07	2019M08	2019M09	2019M10	2019M11	2019M12	2020M01	2020M02	2020M03	2020M04	2020M05	2020M06	2020M07
		Percent Change VS Year Ago												
WSTS Revenue	7.0%	(13.7%)	(14.5%)	(11.6%)	(8.9%)	(7.8%)	1.2%	7.1%	7.4%	6.2%	4.8%	3.4%	7.0%	
WSTS Revenue Americas	8.4%	(27.0%)	(27.2%)	(32.3%)	(16.5%)	(13.0%)	0.3%	21.1%	26.8%	18.1%	29.7%	27.3%	30.1%	
WSTS Revenue Asia-Pacific	8.3%	(11.2%)	(11.8%)	(7.6%)	(6.5%)	(5.8%)	3.9%	5.0%	4.4%	4.3%	3.3%	1.2%	4.0%	
WSTS Revenue China (3MMA)	N/A	(13.9%)	(15.4%)	(13.0%)	(10.3%)	(5.8%)	0.5%	5.1%	7.8%	4.5%	4.5%	3.8%	4.7%	
WSTS Revenue A-P Exc China (3MMA)	N/A	(10.7%)	(8.4%)	(6.3%)	(6.7%)	(7.8%)	(2.3%)	(4.3%)	0.5%	4.6%	3.3%	1.9%	0.4%	
WSTS Revenue Japan	1.7%	(20.1%)	(8.4%)	(5.9%)	(8.4%)	(10.1%)	(5.1%)	0.1%	1.4%	1.9%	(4.1%)	0.8%	(3.0%)	
WSTS Revenue Europe	3.2%	(2.6%)	(6.5%)	(9.1%)	(5.6%)	(8.5%)	(8.2%)	(0.2%)	(2.6%)	(0.6%)	(18.4%)	(22.5%)	(9.2%)	
U.S. IP (SA)Semiconductor and other electronic component	6.3%	1.3%	1.8%	2.7%	5.1%	9.3%	11.0%	9.6%	6.9%	7.0%	3.8%	3.7%	4.7%	5.4%
China Integrated Circuits Output (100 million units)	13.9%	14.2%	22.9%	25.9%	18.0%	21.6%	40.0%	29.1%	29.1%	53.5%	55.6%	30.1%	41.7%	28.8%
Korea IP (SA) Mfg of Semiconductor	17.1%	3.7%	13.1%	9.8%	11.9%	12.6%	15.7%	40.7%	46.0%	45.8%	16.3%	26.9%	23.8%	
Taiwan IP 26: Electronic Parts and Components	4.9%	6.8%	18.3%	3.4%	(1.5%)	8.7%	10.6%	17.7%	22.3%	27.5%	22.9%	23.0%	23.4%	14.4%
Taiwan IP 2611: Integrated Circuits	10.6%	15.1%	21.8%	5.9%	(2.2%)	13.1%	11.2%	30.5%	46.6%	42.5%	33.0%	37.2%	34.3%	
Taiwan Foundry wafer (12 inch & above) (1,000 Pcs)	9.4%	2.0%	10.2%	(1.6%)	(10.4%)	3.1%	14.7%	25.0%	30.5%	31.5%	30.2%	24.5%	29.8%	
Taiwan Foundry wafer (8 inch) (1,000 Pcs)	3.6%	(2.0%)	(12.4%)	(7.7%)	(12.4%)	(3.8%)	(0.1%)	10.1%	18.2%	15.4%	18.7%	19.1%	19.9%	
Taiwan Foundry wafer (6 inch and below) (1,000 Pcs)	2.7%	(13.3%)	(12.3%)	(18.2%)	(42.4%)	(14.2%)	(5.0%)	7.4%	24.9%	27.3%	19.8%	14.2%	17.6%	
Taiwan IP 2613: Packaging and Testing of Semi-conductors	2.4%	9.5%	11.5%	11.1%	8.7%	9.9%	16.6%	19.0%	27.1%	15.3%	13.5%	0.3%	3.5%	
Taiwan DRAM (Dynamic random access memory) (1,000 Pcs)	4.8%	14.6%	15.0%	22.7%	28.7%	34.1%	18.6%	16.2%	27.0%	14.2%	9.2%	6.9%	(6.7%)	
Taiwan Flash disk (Set)	13.3%	38.6%	32.6%	31.1%	7.3%	(22.0%)	10.9%	(15.5%)	13.7%	3.5%	(33.6%)	(46.0%)	(28.1%)	
Japan IP All Integated Circuits	6.0%	(26.2%)	(17.7%)	(22.3%)	(20.2%)	(21.0%)	(16.0%)	2.7%	7.1%	0.6%	(5.3%)	(25.3%)	(1.7%)	
Japan IP Linear ICs 13.6	3.6%	(10.2%)	(6.8%)	(8.0%)	0.9%	3.5%	8.4%	5.6%	15.4%	15.1%	6.1%	0.2%	(8.2%)	
Japan IP MOS ICs (Micro computer) 13.5	(2.6%)	(8.8%)	2.2%	6.0%	5.2%	11.7%	0.3%	(1.6%)	(2.2%)	(1.8%)	(6.6%)	(3.2%)		
Japan IP MOS ICs (Logic) 23.1	(12.1%)	(15.5%)	(17.4%)	0.5%	(13.8%)	(9.7%)	(8.3%)	(15.5%)	(3.3%)	2.0%	(8.0%)	(7.4%)	(13.6%)	
Japan IP MOS ICs (Memory) 84.7	13.4%	(32.4%)	(45.5%)	(52.4%)	(43.9%)	(49.7%)	(46.0%)	0.5%	7.5%	(1.7%)	(4.2%)	(37.4%)	25.6%	
Japan IP MOS ICs (CCD) 32.7	17.7%	8.4%	21.0%	16.2%	13.8%	26.6%	35.6%	17.8%	32.2%	(1.4%)	(5.5%)	(1.9%)		
Japan IP Hybrid ICs 45.5	1.7%	1.6%	(7.2%)	(3.3%)	(18.0%)	(6.1%)	(2.6%)	3.8%	3.6%	1.6%	(12.1%)	(16.4%)	(24.9%)	
SGP IP Semiconductors (3344)	15.4%	8.5%	(28.0%)	(13.7%)	7.4%	(21.2%)	2.7%	0.6%	(24.8%)	11.1%	3.4%	6.4%	22.5%	(0.2%)
Number of Rising Indicators		14	12	12	12	13	19	22	25	25	20	21	19	3
Number of Falling Indicators		15	17	17	17	16	10	7	4	4	9	8	8	1
Net (Rising minus Falling)		(1)	(5)	(5)	(5)	(3)	9	15	21	21	11	13	11	2

Sources: SIA & National statistics databases for China, US, Japan, Taiwan, South Korea, Singapore

Hilltop Economics

Semiconductors out-of-step with economy

Semiconductors (MSI and Revenue) Returned to Growth in 2020

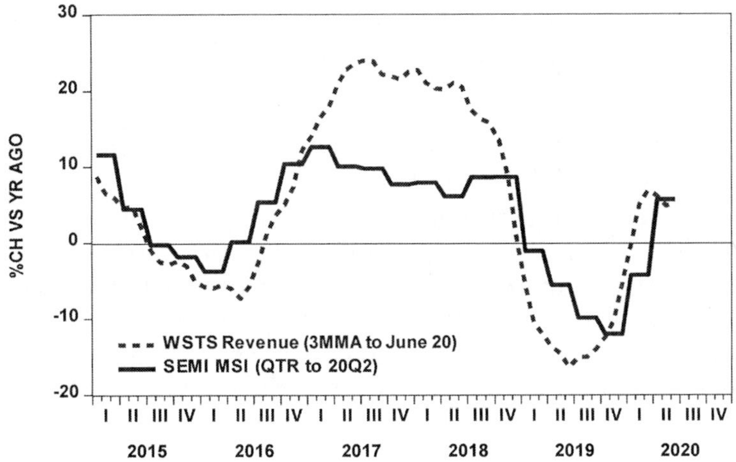

17

Hilltop Economics

Semiconductors out-of-step with economy

A Lot of Uncertainty in Semiconductor Revenue Forecasts

Semiconductor Revenue Forecasts		
Forecaster:	When	2020 % Change
Cowan LRA **	June	6.2
ARIMA (Hilltop Econ X-13)**	August	3.9
WSTS	June	3.3
IC Insights	June	3.0
Omdia	May	2.5
VLSI	June	0.7
Semico	May	-0.2
IDC	May	-4.2
Semiconductor Intelligence	May	-6.0
OECD LEI (Hilltop Econ "V" economy)	July	-8.3
Gartner	July	-9.6
McKinsey (midpoint)	April	-10.0
IBS	July	-10.4
Consensus Revenue Forecast	July	-1.8%
Standard Deviation		+/- 5.5%

** Time series or trend-only methods

18

Hilltop Economics

Big Change in Spending Behavior in the Pandemic Recession

- Normal Behavior in Recessions:
  - Consumers cut spending on discretionary items – including technology goods
  - Businesses delay investment in new PCs, equipment, etc. to conserve cash and wait for capacity utilization to recover
- Pandemic Recession Behavior:
  - Lock-downs, work-from-home, remote education all contribute to a spike in spending, especially where incomes have been supported by government assistance programs
  - Real GDP severe declines come from declines in services (travel & entertainment, discretionary activities including elective medical, any activity requiring people associating in close contact)

Hilltop Economics

Hilltop Economics

Behavioral Change: Source of Semiconductor Uncertainty

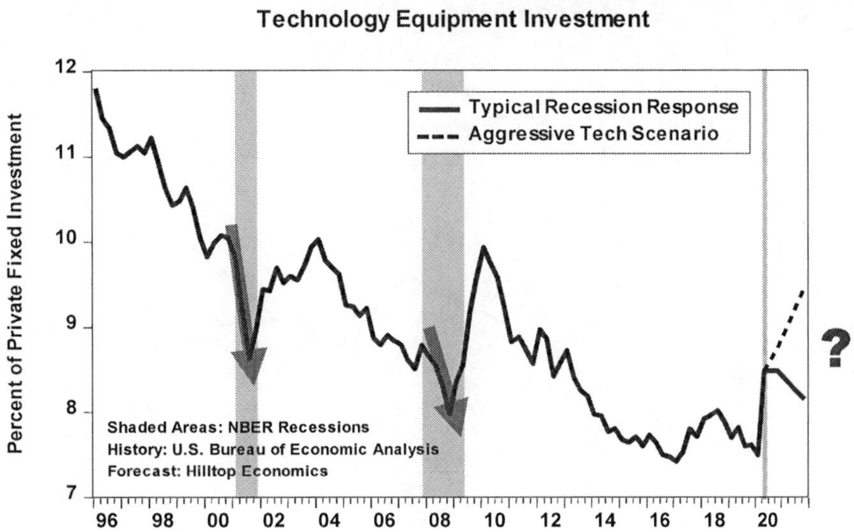

Hilltop Economics

Scenarios to manage through uncertainty

Scenario Overviews (August 29, 2020)

- **August Consensus Forecast: a modified "V"-shaped global recession**
  - Severe Covid-19 impact, but successfully managed
  - V-shaped global recession extremely disruptive in the short run but healthy economies rebound over the 2021-2022 period as major economies stimulate with large fiscal support policies.
  - Tech-related investment reacts in modified traditional way: initial boost, then follows collapse/recovery in final demand
- **Aggressive Tech Investment in a modified "V"-shaped global recession**
  - Same economic/pandemic backdrop as Consensus Forecast
  - Tech-related investment prioritized over other investment categories throughout the recession and then into recovery, driven by aggressive investment.
- **Extended Pandemic – A flat "U" or "L" shaped global recession**
  - Severe Covid-19 impact (e.g. ~9-12 months of severe impact, delayed or less-effective vaccines, weakly successful treatment protocols).
  - The pandemic produces permanent structural changes in the global economy.
  - Extended and severe disruptions in demand. Fiscal and monetary responses are ineffective or counter-productive, and leave economies too weakened to recover quickly. Tech investment follows traditional behavior in recession.

Hilltop Economics

Scenarios to manage through uncertainty

Scenario Real GDP Levels

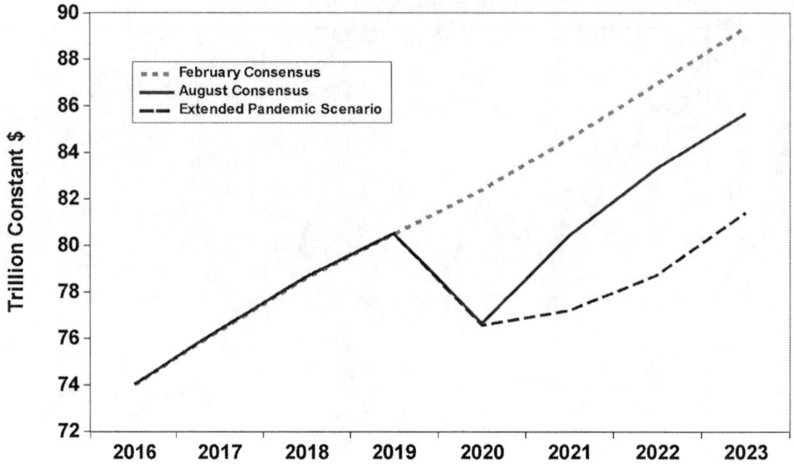

23

Hilltop Economics

Scenarios based on economics & behavioral assumptions

Scenarios Give Wide Range of MSI Outcomes

Year 2018 MSI Exceeded - Aggressive Tech: 2022; Consensus-Based: 2023; Extended Pandemic: 2024

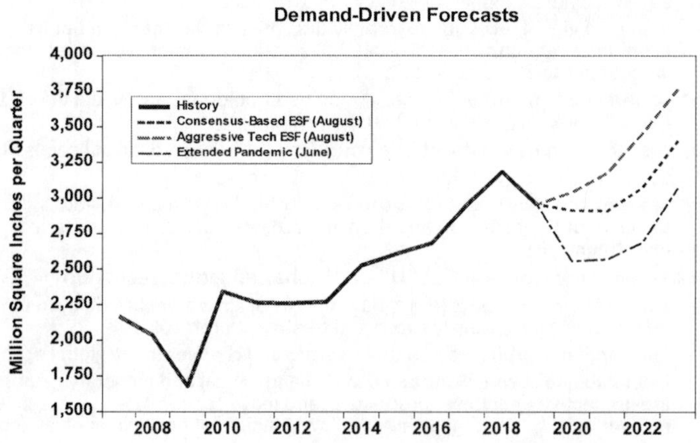

24

Hilltop Economics

Summary

- Downstream: Real GDP and components in deepest recession since 1930s
  - Outlook for recovery modest
  - Pandemic causing monthly forecast revisions: the greater the pandemic impact, the sharper the economic forecast downgrades
  - The pandemic ultimately will drive the economics
- Upstream: most markets growing
  - Initial surge from work-from-home; educate-from-home policies
  - Trade sanctions causing another disturbance to normal market function
- Semiconductors
  - So far, healthy recovery from 2019 downturn continues
  - Pandemic changed behavior relative to past recessions. Outlook depends in part on whether behavior reverts to "typical" in the months ahead.
  - Range of possible outcomes necessitates flexibility, scenarios and constant willingness to revise forecasts as new data become available

Entegris

Future Post-CMP Cleaning Challenges

Michael White*, Daniela White, Atanu Das,
Thomas Parson and Michael Owens,
August 31, 2020

ENTEGRIS PROPRIETARY AND CONFIDENTIAL

Agenda

Entegris

- Market trends in PCMP
- Cleaning Silica from dielectrics
- Tribology in cleaner-brush technology
- Cleaning CeO_2 from dielectrics
- The future of PCMP cleaning

2 | CONFIDENTIAL

Post-CMP Cleaning Challenges

Entegris

- New materials/film types
  - Diversification of dielectric materials
  - New barrier materials
- Feature size decreasing
  - 10 nm ubiquitous
  - Working on 3-5 nm for most customers
- Increasingly complex CMP slurries
  - New particle types
  - Small particles
  - Advanced organic additives
- Defect detection thresholds decreasing
  - Current state of the art → 18 nm
- Environmental laws/customer EHS more stringent
  - Country to country variations
  - Customer EHS requirements vary
  - Tetramethylammonium hydroxide increasingly forbidden

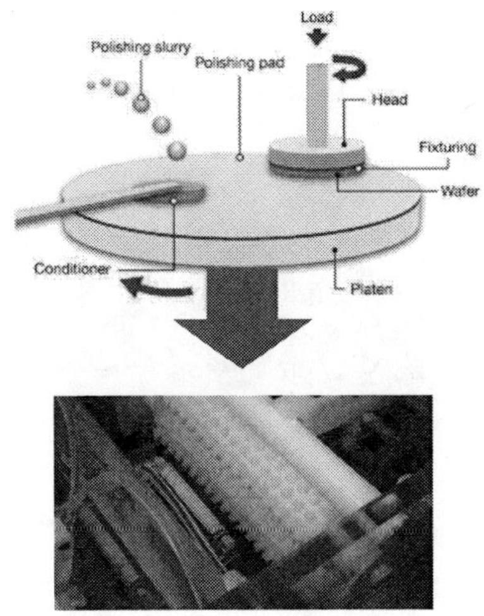

Graphic source: http://www.azom.com/article.aspx?ArticleID=12537

PCMP Roadmap Challenges

Entegris

2016	2018	2020	2022	2024
10 nm	7 nm	5nm	3 nm	

Copper / Tantalum

Challenges:
Organic residue
Silica defects
Low k Defectivity
Cu corrosion
Cu Ring Scratches

Cobalt Liner
Additional Challenges
Cu-Co galvanic corrosion

Ruthenium ?

Ru Liner and Ru Bulk
Additional Challenges:
Ru metal or RuO_2 residues

?

Mn
Pt Group (Rh, Ir, Ru, Os, Pt, Pd)
Binary compounds (MoN, Re_2C, ...)

Tungsten

Cobalt Interconnect
Additional Challenges
Cobalt metal or cobalt complex ppts

Challenges:
Organic residue (surfactant on nitride)
Silica defects
Fe and Ti residues
Low recess

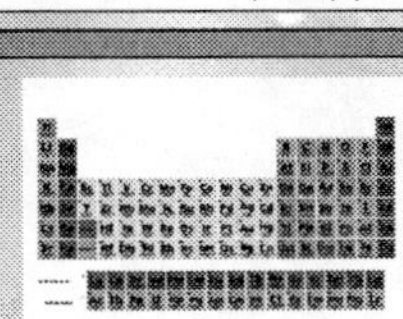

Molybdenum

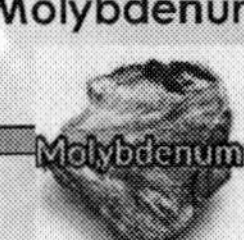

Mo Additional Challenges:
Higher etch rates than W

Novel Dielectrics

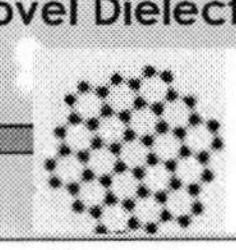

Silsesquioxane
CVD C
New Low k
Other metal oxides

Graphic source: Wikipedia

The Big CTC Bang in the Formulated PCMP Universe

Entegris

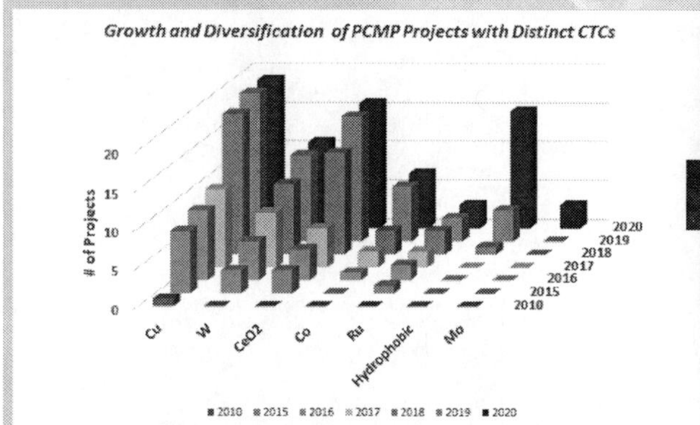

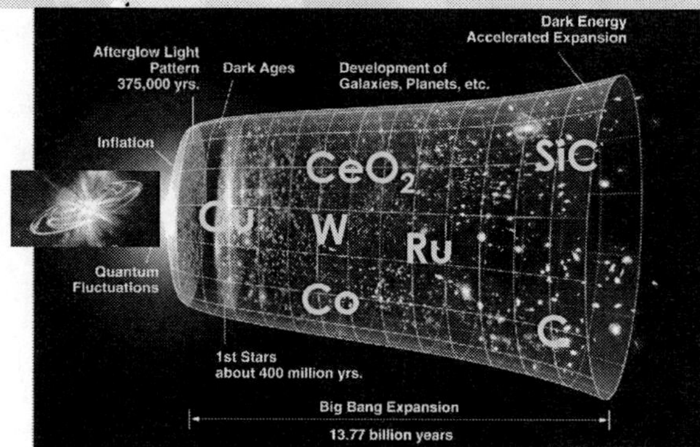

Graphic source: Wikipedia and interestingemgineering.com

# of projects is defined as a potential separate major customer projects with CTCs that are different → novel formulations required

Types of organic Residue Encountered in Post-CMP Cleaning

- Corrosion-inhibitor – metal complexes (i.e., Cu-BTA)
- Surfactant residues
- Interactions between slurry additives and cleaning additives
- Pad debris (polyurethane – hydrophobic or hydrophilic)
- Brush debris (crosslinked PVA – hydrophilic)
- Plating additives
- Filter or tubing shedding/residues
- Biological debris (bacterial, skin cells, etc.)

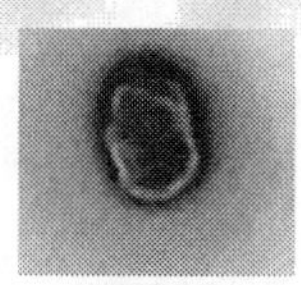

Pad or brush debris
EDX = C, N, O → polyurethane pad
EDX C, O only → brush

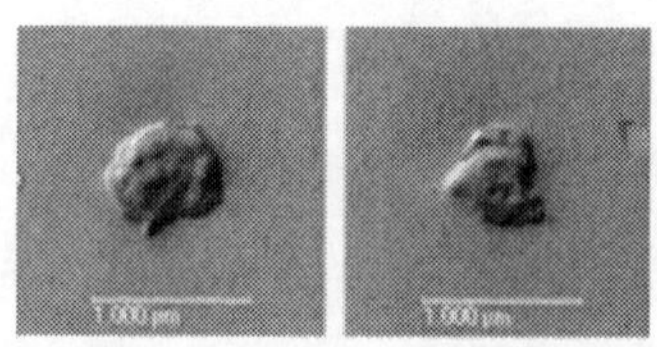

Precipitated organic such as from the interaction of cleaner with slurry components

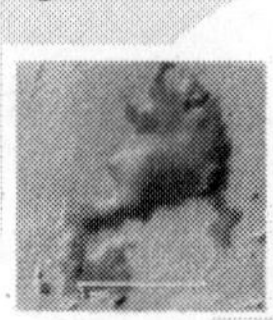

Soluble organic
or
surfactant residue

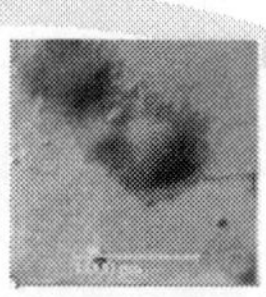

Advanced defect metrology has enabled detection of smaller defect sizes

Entegris

Process Can Often Play a Large Role in Defectivity- Impact of the Brush Gap

Brush cleaning:
physical + chemical cleaning

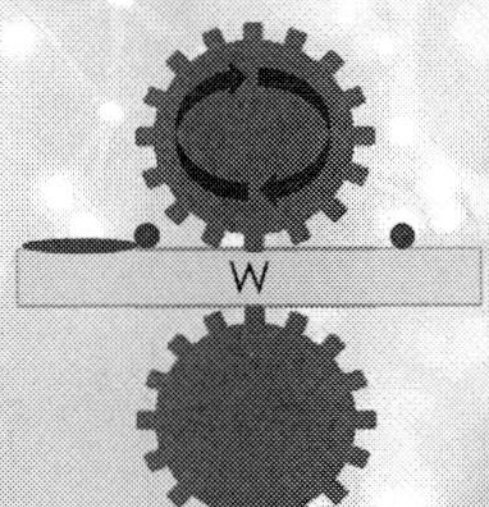

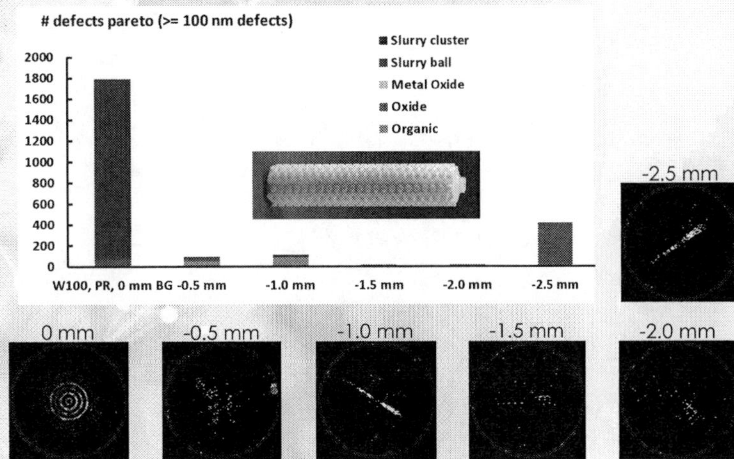

Brush Pressure:

- Too low: insufficient physical cleaning
- Too high: particles stick to PVA brushes and lead to brush marks

7 Entegris

Diversity in Particle Types Leads to Various Cleaning Challenges in Silica Cleaning

Entegris

- **Fumed silica (older ILD, W slurries)**
- **Colloidal silica**
  - **Controlled growth from "silicic acid"**
  - **High purity particles via Stober process – controlled hydrolysis of Tetraalkylorthosilicate**
- **Surface functionalized silica**
  - **Anionic functionality (carboxyl or sulfonic acid)**
  - **Cationic (amines, quaternary ammonium)**

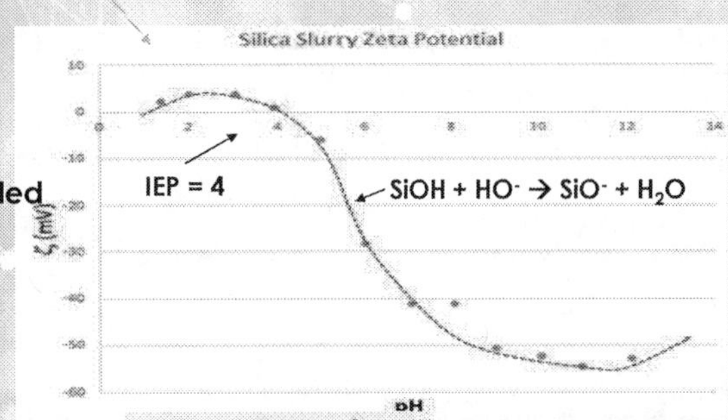

Charge attraction/repulsion, H-bonding and hydrophobic forces lead to large interactions between the Silica particles, the brush and the surfaces

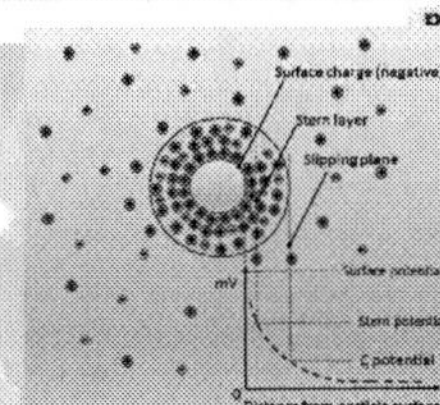

Zeta potential
$\zeta = 4\pi\gamma(v/E)/\varepsilon$

8

Theoretical Tribology Background

Entegris

$COF = F_{shear} / F_{normal}$

Ref: Philipposian, A. et al, J. El. Chem. Soc. 151(7) G456-460 (2004)

Sommerfeld number

$So = \mu \times U / p \times h_{eff}$

- μ is the viscosity of the cleaning fluid,
- U is the relative brush-wafer velocity,
- p is the applied brush pressure on the wafer,
- h_{eff} is the effective fluid film thickness

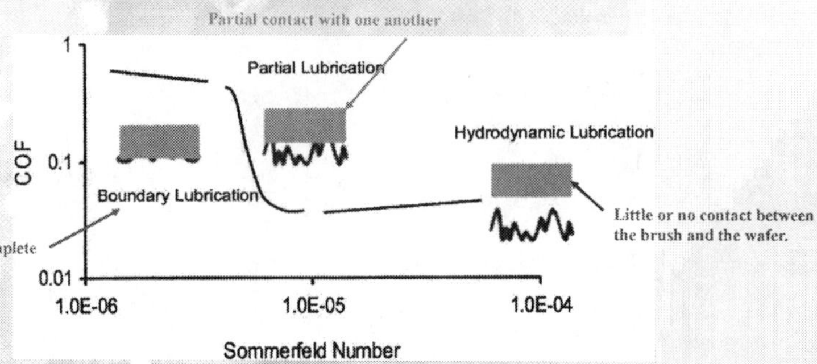

The Stribeck curve is based on the Sommerfeld number

- $F_{shear,}$ the shear force is determined at a frequency of 1000 times per second from the force transducer attached to the sliding friction table;
- $F_{normal,}$ the normal force is found experimentally using the pressure sensor technology;

In-Situ Brush Tribology

Entegris

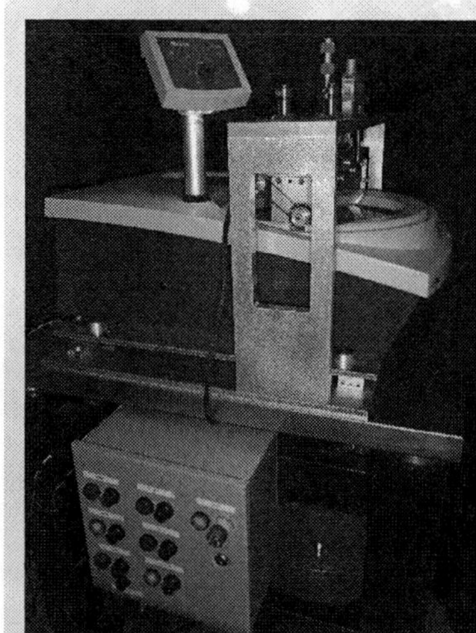

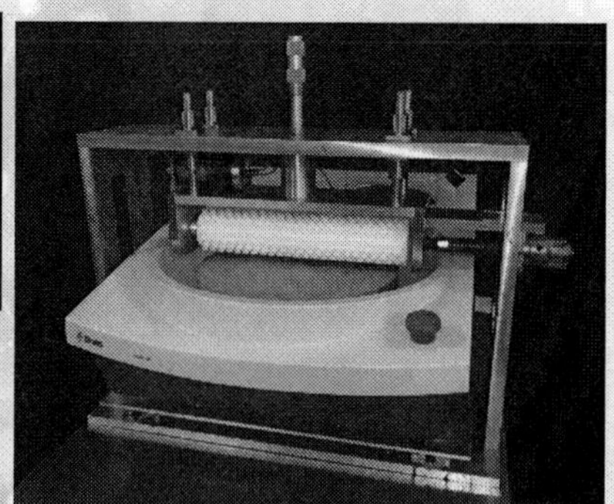

Capabilities:
1. Measure Friction under dynamic conditions
2. Brush tension and speed controllable to model customer processes

Friction measured n Araca tool

10 | CONFIDENTIAL

AG- W Series Show Significantly Lower Defects than Commodity Cleaner

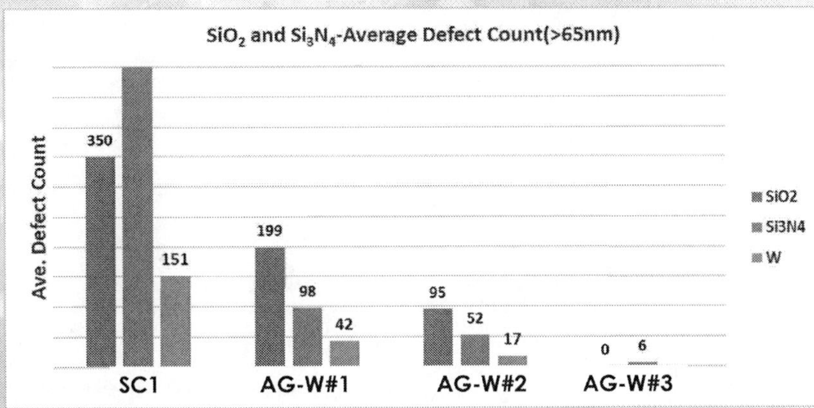

- **Defectivity determined at IMEC on 300mm Reflexion with SP3 at 65 nm Threshold**

11 | CONFIDENTIAL

Brush Friction/Tribology and Flow Depends on the pH and Cleaning Additives

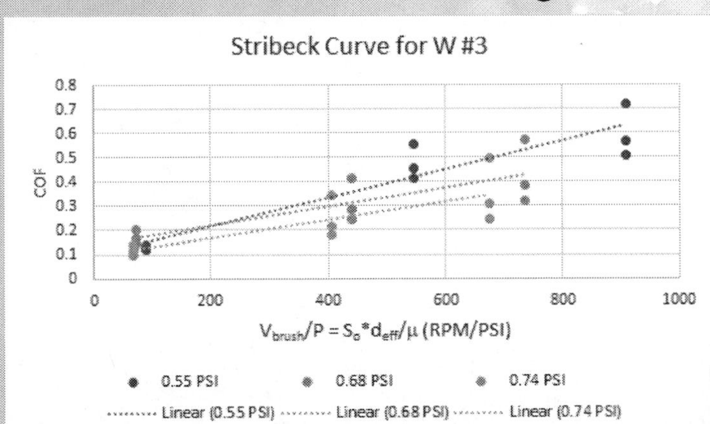

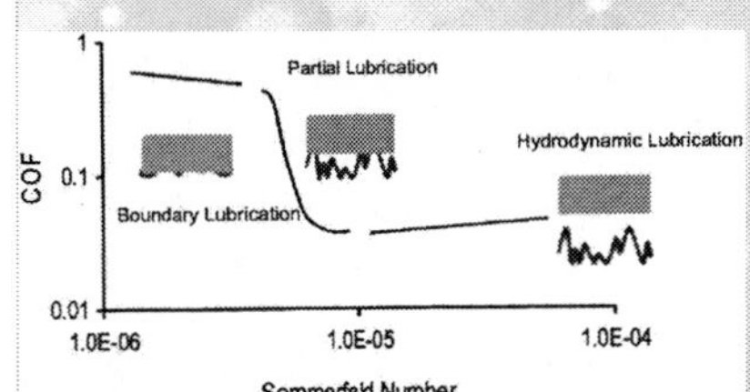

Ref: Philipposian, A. et al, J. El. Chem. Soc. 151(7) G456-460 (2004)

AG- W#2 is in the hydrodynamic lubrication regime

12

Correlation Between IMEC Defectivity and COF

Entegris

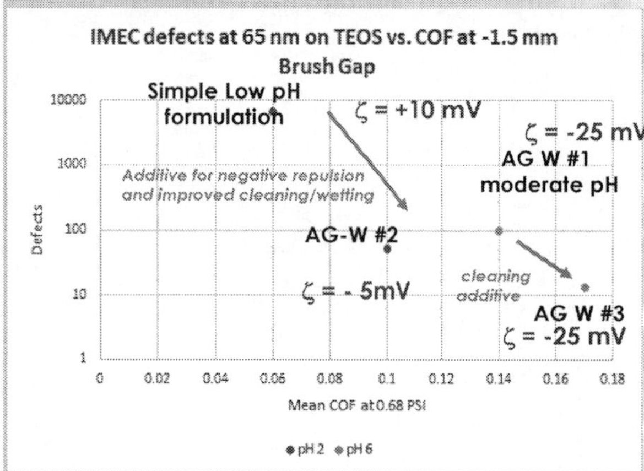

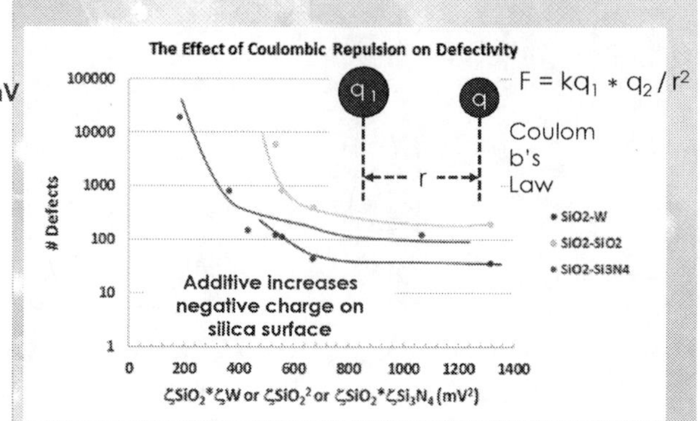

Increasing brush friction and increased charge repulsion both impact cleaning

13 | CONFIDENTIAL

Requirements for a Post-CMP Ceria Cleaner

Post-CMP Cleaning Formulations

- Many different ceria types
  - Calcined
  - Precipitated/colloidal
- Organic additives vary depending on goals
  - Rate additives (ILD)
  - Polymeric or small molecule selectivity additives
    - Stop on nitride or poly
    - Reduce feature size dependence

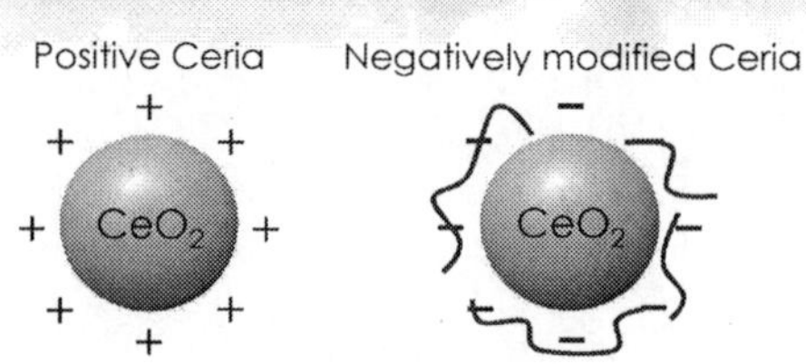

Current industrial POR (commodity clean): inefficient and environmentally unfriendly

- DHF
- SC-1
- SC-1 + DHF
- SPM (H_2O_2 + H_2SO_4)
- TMAH + SC-1

- Highly toxic
- Unformulated
- Environmentally unfriendly
- Damage to dielectric

Entegris AG Ce-XXXX formulations

- Improved particle and metal removal
- Replace toxic commodity cleaners
- Environmentally friendly
- No damage to dielectric surfaces
- Ce residue post-CMP cleaning < 10^{10} atoms/cm^2

14 ENTEGRIS PROPRIETARY AND CONFIDENTIAL – INTERNAL

Entegris

Ceria PCMP Challenges

Entegris

- Removal of Ceria particles from surface
  - CTC – as low as possible

  Challenge: Calcined vs. colloidal ceria abrasive shows different degrees of interaction with the surface

- Removal of Ceria-atom (Ceria-atom dissolution)
  - Ce-atom/cm^2 = 10^9 atm /cm^2

  Challenge: CeO_2 dissolution requires high activation energy; slow kinetics

- Etch rate
  - Low W-etch rate (some customers)
  - Low PETEOS and SiN

  Challenge: maintain low substrate etch rate

- No organic residue

  Challenge: additive chemistry to eliminate organic residue

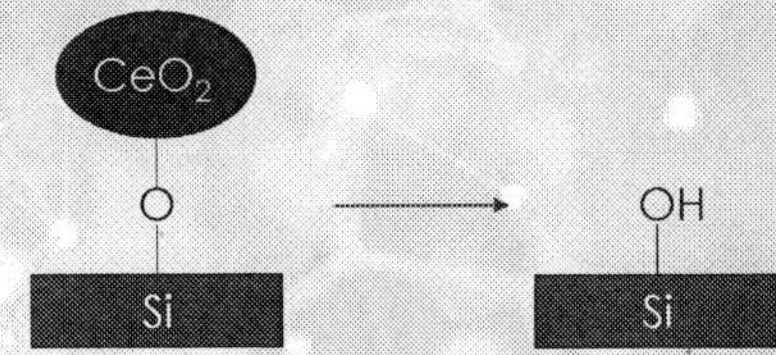

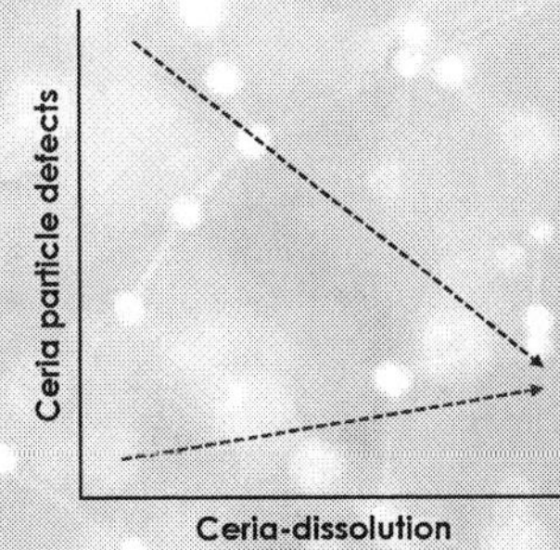

15 | CONFIDENTIAL

Particle Diversity in Ceria Cleaning

Entegris

- **Cationic surface at low pH**
- **Adsorbed additives for selectivity can modify surface properties**
- **Mixed Ce(III) and Ce(IV) surface oxidation states –variable depending upon calcining conditions**
- **Morphology variable depending upon calcining conditions**
- **Market driving towards smaller mean particle size for reduced micro-scratches**
- **Smallest particles – highest radius of curvature (energy to remove is higher)and least shear force under brushes**
- **Other cationic particles are also diverse**
  - **alumina (alpha vs. gamma vs. hydrated)**
  - **Titania (rutile vs. anatase) , zirconia,...)**

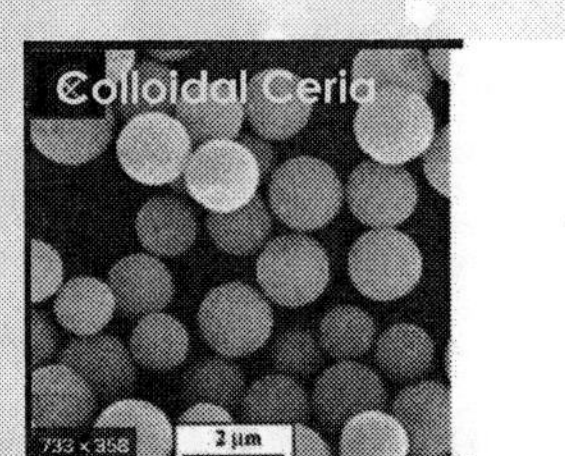

Ref: Andreescu, D. *Coll. and Surf A*, v291 pp. 93-100

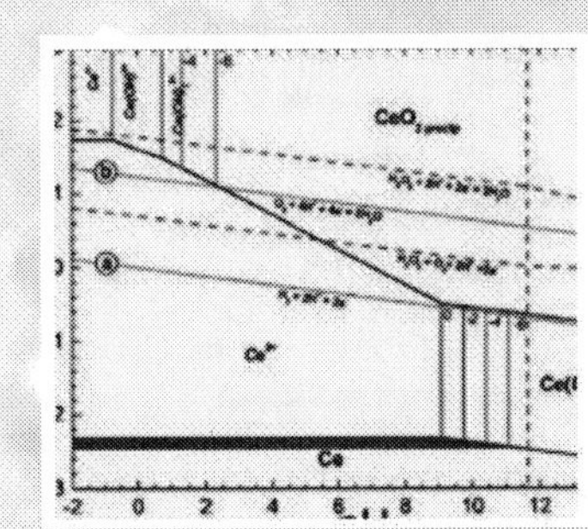

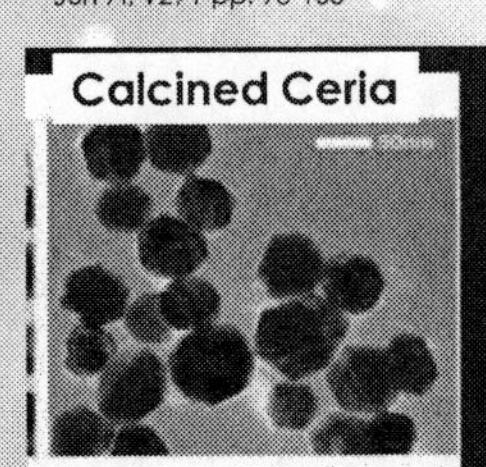

Ref: Kang, et al: ICPT 2012 6353752

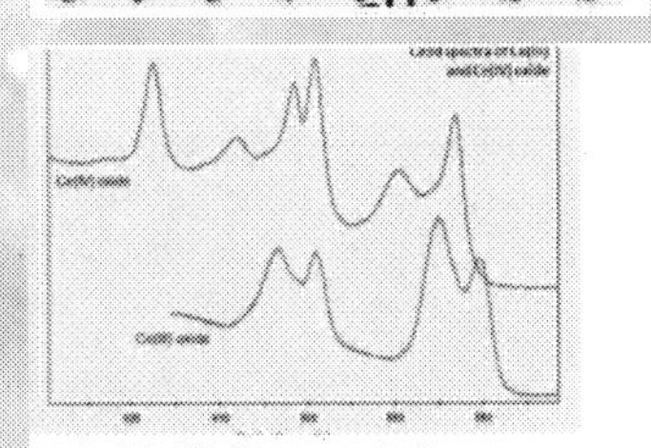

Thermo Scientific XPS: Knowledge Base
xpssimplified.com

16

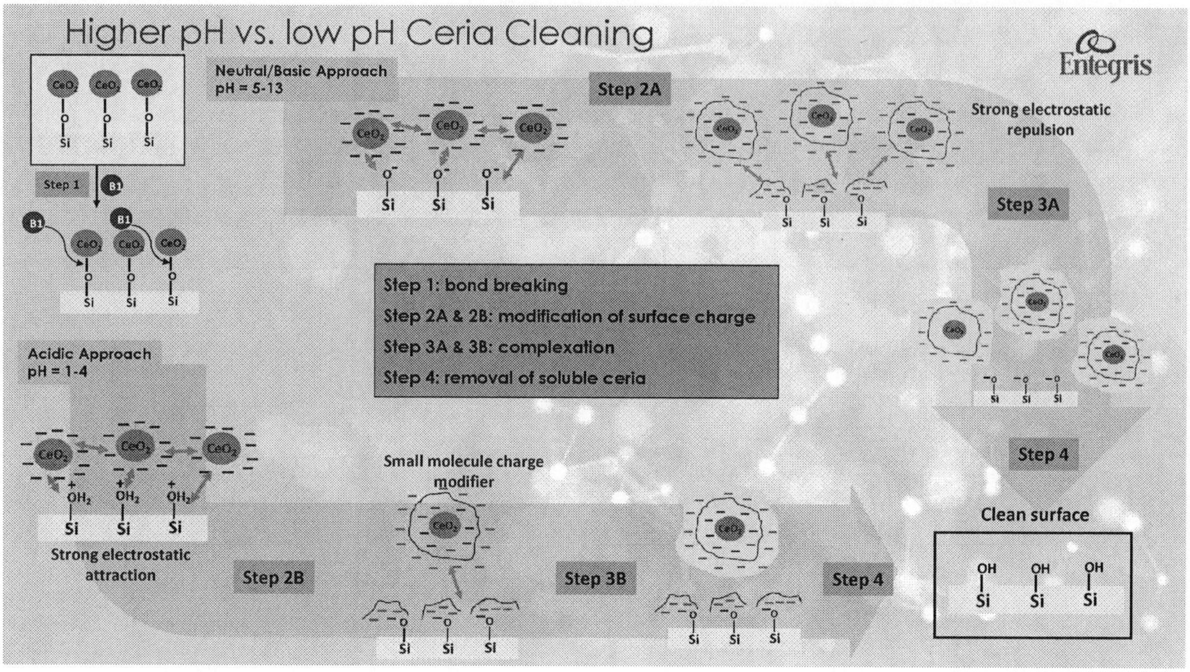
Higher pH vs. low pH Ceria Cleaning
Entegris
Neutral/Basic Approach
pH = 5-13
Step 2A
Strong electrostatic repulsion
Step 1
Step 3A
Step 1: bond breaking
Step 2A & 2B: modification of surface charge
Step 3A & 3B: complexation
Step 4: removal of soluble ceria
Acidic Approach
pH = 1-4
Step 4
Small molecule charge modifier
Clean surface
Strong electrostatic attraction
Step 2B
Step 3B
Step 4

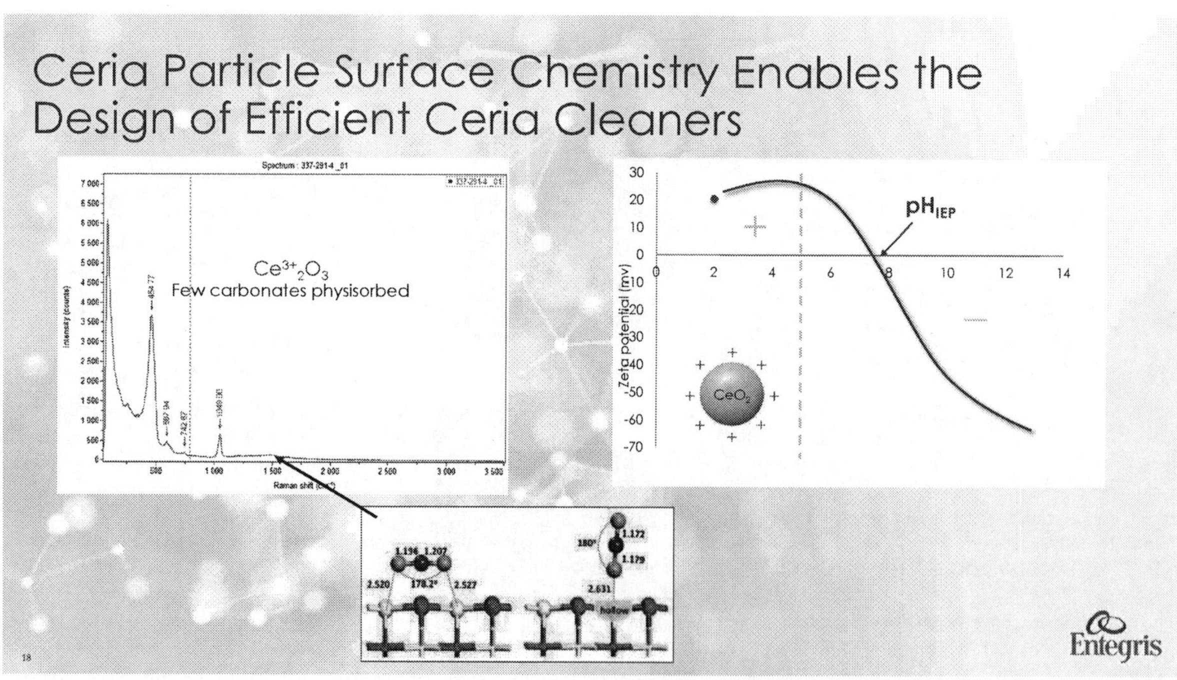
Ceria Particle Surface Chemistry Enables the Design of Efficient Ceria Cleaners
Ce3+2O3
Few carbonates physisorbed
pHIEP
Zeta potential (mv)
Entegris
18

FTIR IR Spectra of Cerium Silicate Shows the Effectiveness Bond Breaking Agent

Entegris

Si – O – Ce solid sample + CeXXXX sample ⟶ Solid material isolated + filtrate isolated

IR of sample before and after treatment

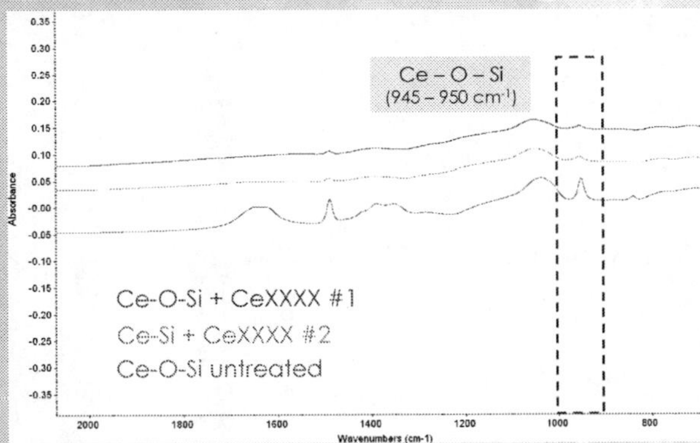

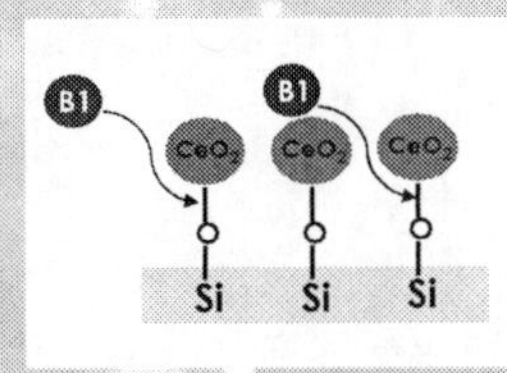

Relative Ce – O – Si intensity is greatly reduced with CeXXXX

19

Advanced Products Have Higher CeO_2 Dissolution Efficiency

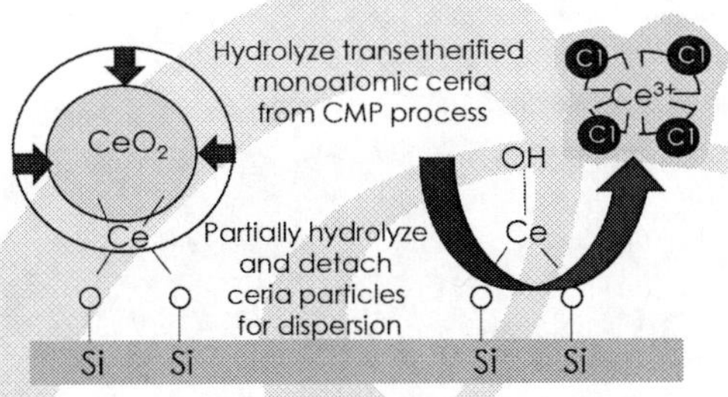

Ceria-Slurry (CeO_2) Dissolution

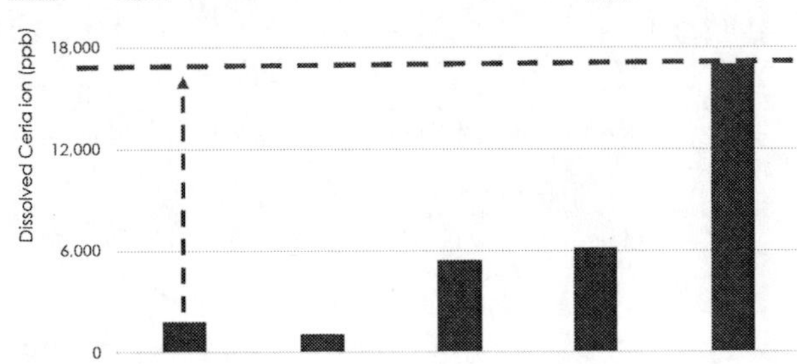

Why is increased ceria dissolution desirable?

Entegris

20

ENTEGRIS PROPRIETARY AND CONFIDENTIAL – INTERNAL

Full Wafer Cleaning Data (AMAT Mirra / OnTrak SP1) Show Defectivity Improvement Over Commodities

Entegris

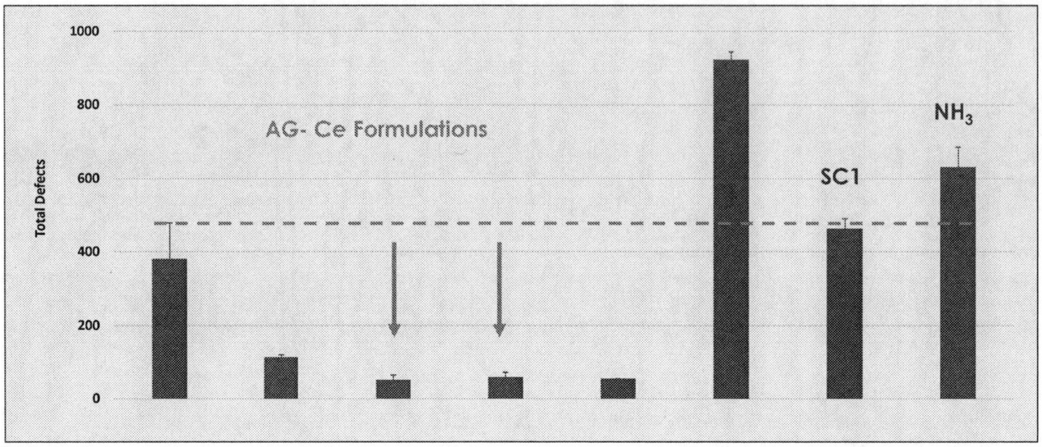

21 | CONFIDENTIAL

Conclusions

Entegris

Entegris believes that advanced metrology and better simulations of customer processes are the key to designing more effective PCMP cleaners

- *In-situ* electrochemistry (Tafel, impedance) under brushing conditions
- *In-situ* brush tribology and fluid flow characteristics
- Model reactions such as cerium silicate bond-breaking studies
- Measurement of particle-wafer interaction energies
- Spectroscopic analysis of surfaces, particles and side reactions
  (UV-Vis, FTIR, TOF-SIMS, XPS, Raman, 1H/13C/multinuclear NMR)

Proper mechanistic studies have led to superior defectivity and advanced node patterned wafer corrosion performance

22

Thank you!

Entegris

Q&A

Appendix

Wet Etch Challenges and Innovation Driving Nano-Electronics Scaling

Nabil Mistkawi, Ph.D.
Senior Staff Engineer/Technologist

SAMSUNG
AUSTIN SEMICONDUCTOR

August 31st, 2020

SPCC Business of Cleans

Outline

- IRDS Requirements & Expectations
- Cost of Contamination and Excursions
- Lines of Defense
- Beyond 10nm – Scaling in the New Era
- Suppliers Call to Action
- Summary

SAMSUNG
AUSTIN SEMICONDUCTOR

IRDS Roadmap with Respect to Contamination

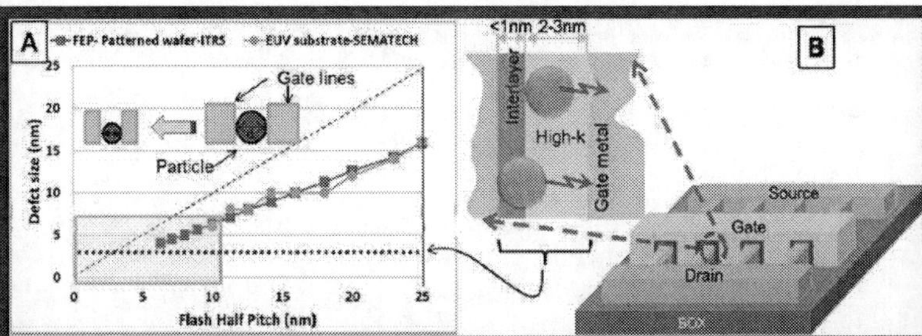

As we continue to march towards smaller dimensions, sub 10nm particles/defects will have profound impact on Yield and device reliability performance

A. Rastegar, et.al, SPIE 9048, EUV Lithography V, 90480P (2014)

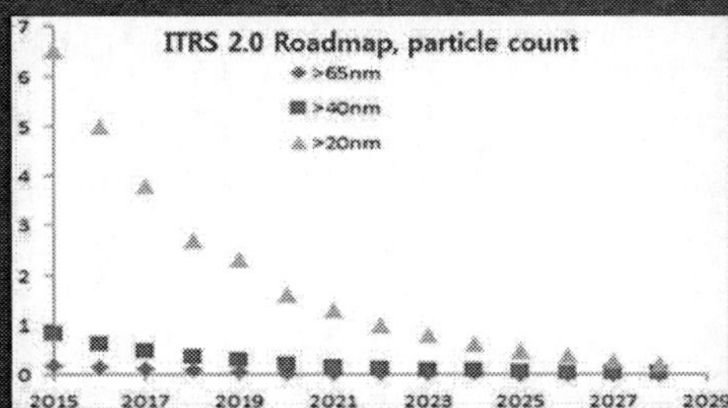

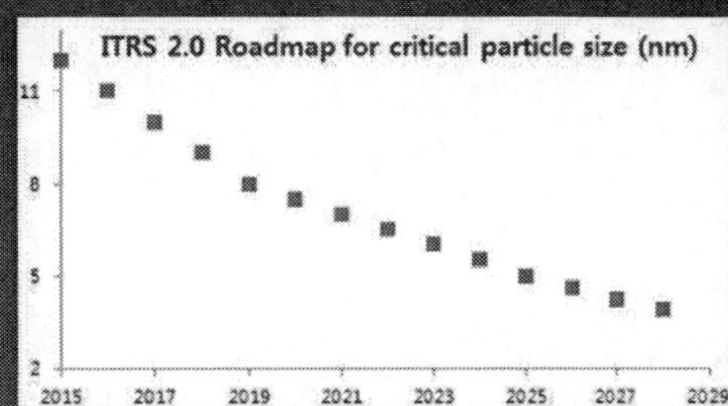

IRDS Particle count and size projections create challenges and opportunities for the industry

Cost of Contamination/Excursions

Possible Sources of Particles/Contamination

- Corrosion products → Metal-rich residue
- Fab environment/facilities
- Contamination from the etching/deposition chamber
- Byproduct from plasma/vapor phase etching
- Incoming Particles → Chemical supply/raw material
- Surface properties – Hydrophilic vs. hydrophobic
- Zeta potential difference between particle and wafer surface

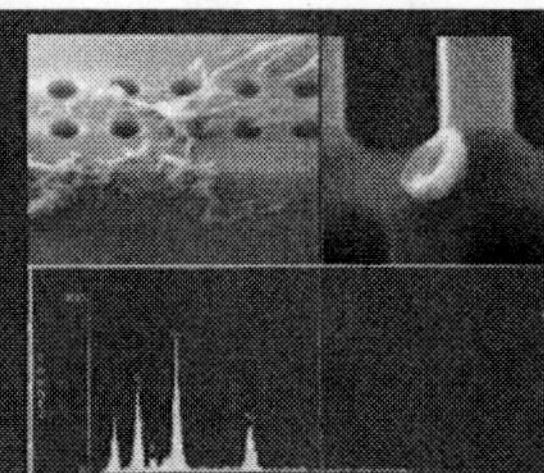

Source: Nabil Mistkawi, Dissertation

Determining the source/root cause is frequently challenging but extremely important to identify a remedy

- **Contaminated raw material effects finished goods quality**
- **Customers cannot afford excursions: one failure can have a significant financial impact**

Early Detection Lowers Impact to Finished products

Lines of Defense

Purification | Filtration and Packaging | Transportation /Transfer

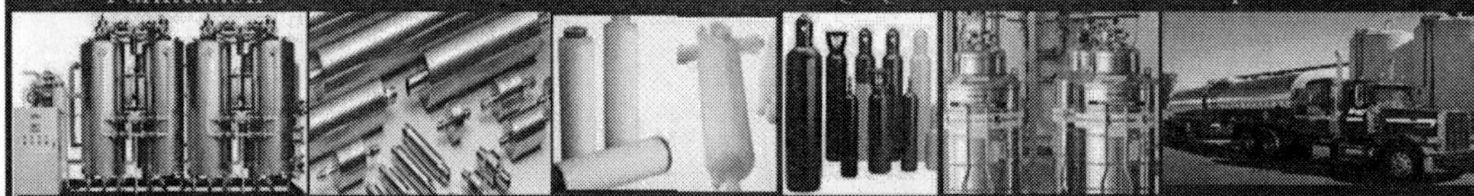

- Raw material sources, purification, and filtration are crucial to semiconductor processing.
- Much stricter protocols (CoA) and further innovations in improving purity and quality control is a key.
- Filters can't introduce contaminants. They must remove and retain impurities without reacting with process liquids and gases
- Inline and continuous monitoring is vital in catching issues before they occur
- Identifying potential failures and their impact is critical → FMEA
- Collaboration/joint efforts with materials suppliers is needed to drive further improvements.

Beyond 10nm

- Non Traditional Scaling – 3D architecture, design innovations, and process sensitivity are increasing
  - ➢ Sub 10nm particles are now significant → Enhanced Purity is crucial
  - ➢ Filtration Technology must keep up with scaling to meet the paradigm shift in defect tolerance
  - ➢ Trace levels of contamination can now have severe impact on yield and device reliability
- Understand particle metrology gap → On blanket wafers vs. in-situ vs. In-line
- Advanced metrology and analytical methods are critical for Defect Identification and Characterization
- **Decision making will be difficult/flawed without stable and capable metrology**

SAMSUNG
AUSTIN SEMICONDUCTOR

Device Technology Roadmap

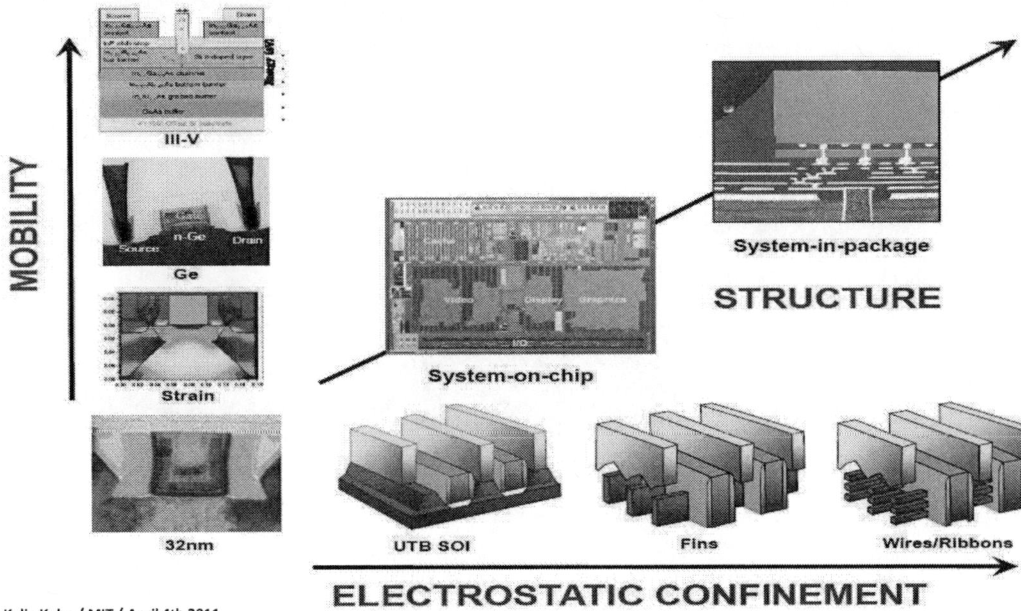

Kelin Kuhn / MIT / April 4th 2011

Fin Collapse

Shear Stress Force

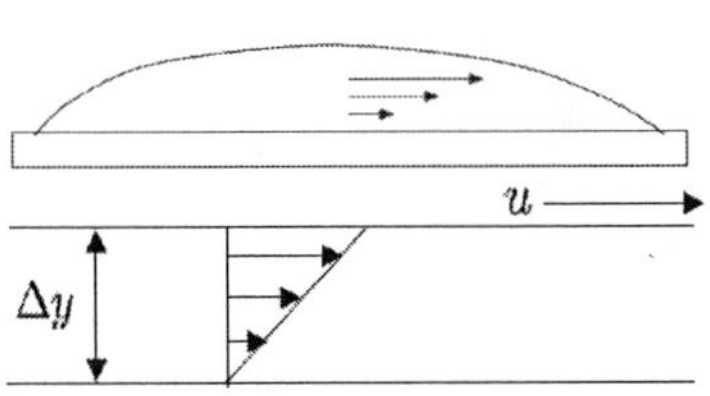

The Shear Stress $\tau(y) = \mu \frac{\partial u}{\partial y} = \rho\omega^2 r$

where
r = distance from center
μ is the dynamic viscosity of the fluid;
u is the velocity of the fluid along the boundary
y is the height above the boundary.
ρ = density, ω = angular speed

Note the direct relationship between the shear/stress force and density/viscosity

Unbalanced Capillary forces During the drying step

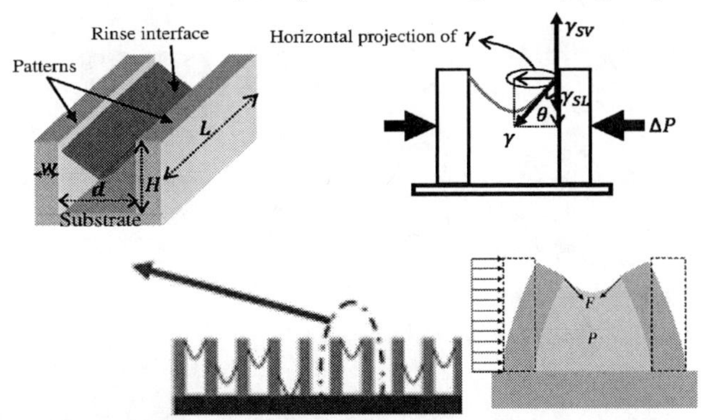

The max stress is described according to:

$$\sigma = \frac{6 \cdot \gamma \cdot \cos\theta}{D} \cdot \left(\frac{H}{W}\right)^2$$

Where γ is the surface tension of the liquid, θ is the contact angle, D is the distance between features, and H and W are the height and width of features respectively.

SAMSUNG
AUSTIN SEMICONDUCTOR

Fin Collapse Prevention

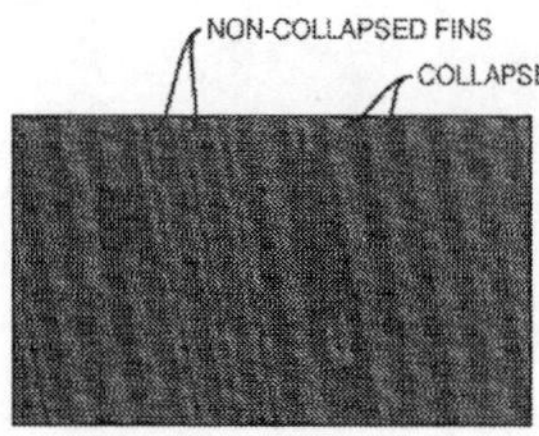

100% no collapse is required

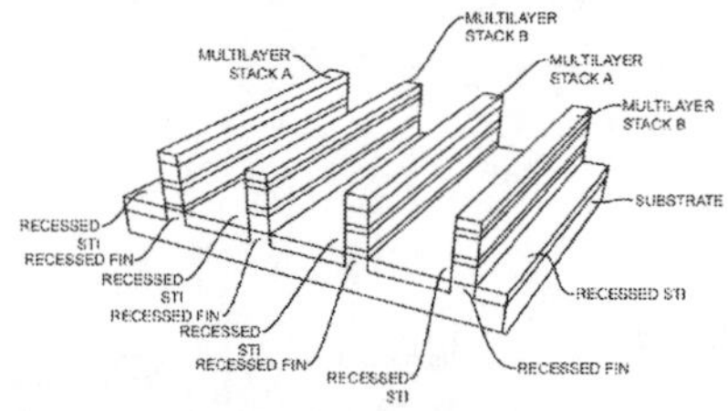

Fin complexity: Heterogeneous multi stack or encapsulated in oxide or nitride films

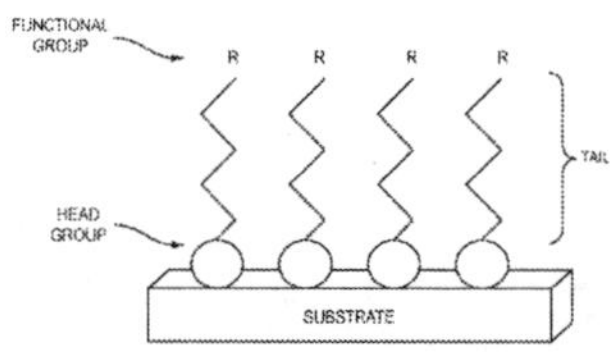

Self assembled monolayer to control surface energy and contact angle

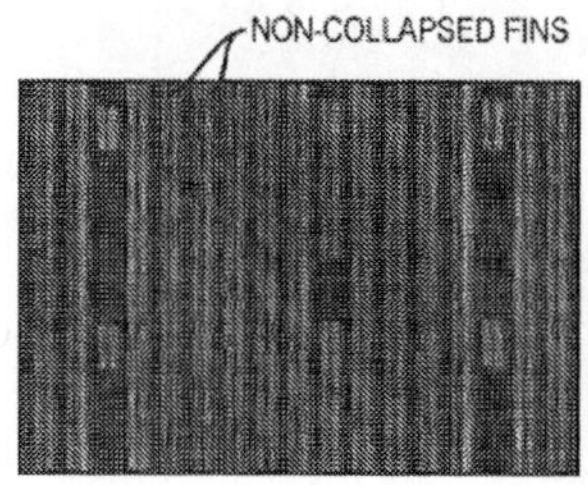

Multistep rinsing including surface derivatization and modification is Needed

Source: Patent **WO 2018/063277 A1**
Inventor: Nabil Mistkawi

Future Device Architecture - GAA

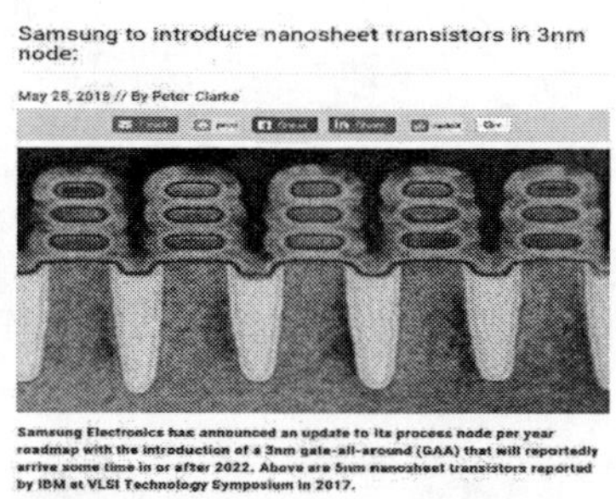

- Semiconductor industry has been able to successfully fabricate Si Nanowires for a decade.
- Fabrication of SiGe Nanowires has yet to be demonstrated.

Gate-all-around (GAA) proposed fo
r
technology nodes beyond 7nm

IMEC & Others - No Progress/Breakthrough on SiGe Nanowires

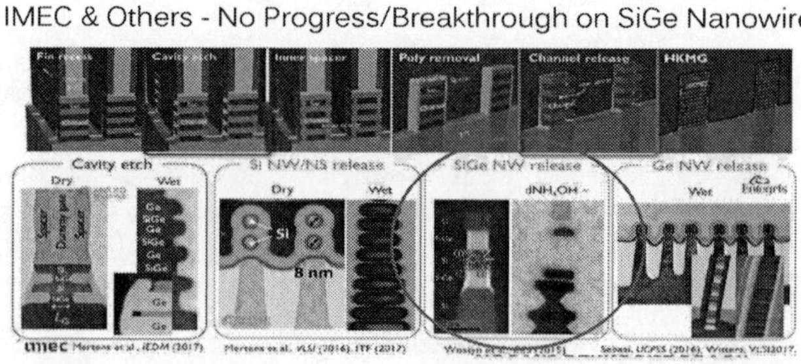

Fabrication of Nanowire Channel Transistor

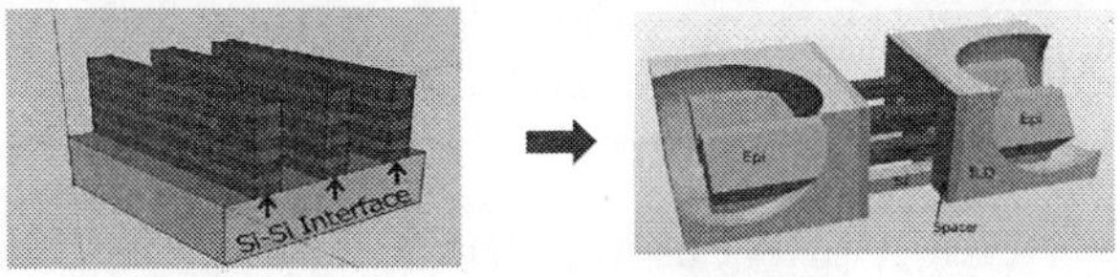

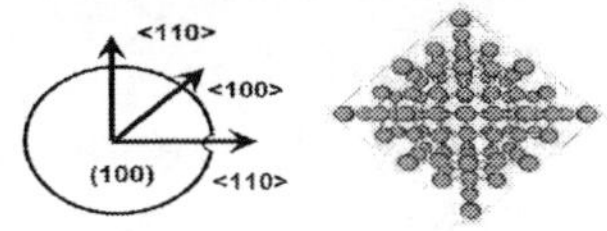

- Must Etch Silicon only in the <110> plane only
- Must Be compatible with SiGe

Selective Etch

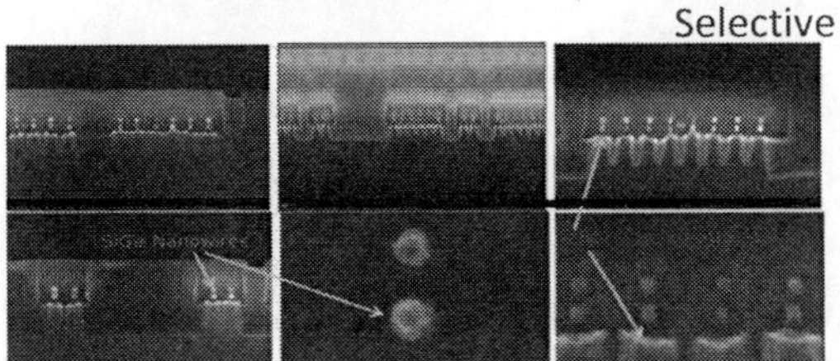

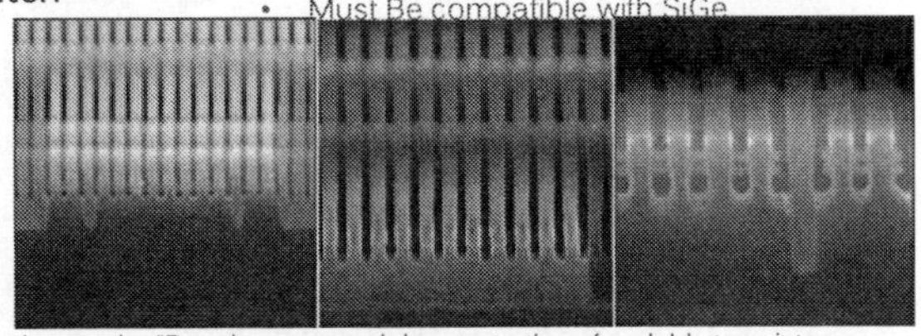

Intel's Achievement Award (Intel's highest technical award) for contributions to the "Development and demonstration of scalable transistor arc hitecture for 7nm and beyond", 2014.

Non-Selective Etch

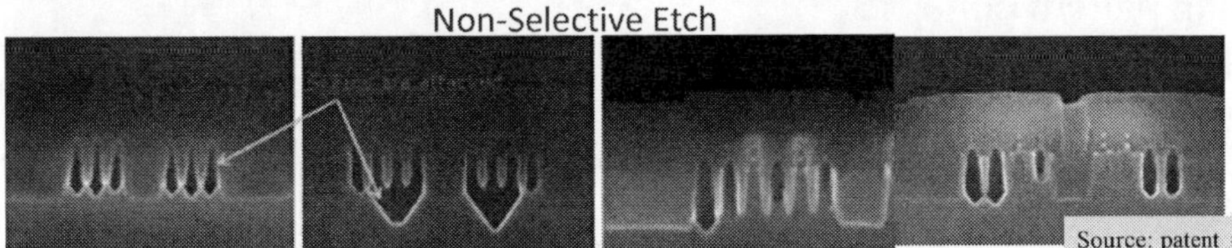

Source: patent #WO2018063300A1
Inventor: Nabil Mistkawi

Requirements & Expectations - Suppliers Call to Action

- **As we continue scaling, stringent customers requirement, tighter defect control/process margins, increase in gas related process steps and complexity. It requires:**
  - High Purity gases and improved filtration technology capable of capturing sub 10nm particles to prevent excursions
  - Change point management and defense line enhancements
  - Tools in the fab must always be up and running - No interruptions in supply chain → gases/raw materials
  - Invest proactively to secure the supply chain → Increase efficiency and Reduce total cost of ownership
  - Further enhancements in quality and reliability to eliminate contamination and improve purity
    - Provides better etch/deposition selectivity
    - Reduce variability at all levels of supply chain and alleviate supply chain risks
  - Swift excursion recovery including root cause identification, containment, and data transparency
  - Create environmentally friendly, innovative solutions, and apply automotive quality approaches… FMEA

Defect reduction, continuous yield improvements and keeping cost per transistor low is the key to the industry's success!!

Summary

- Beyond 10nm process technology nodes would require ultra pure water chemicals and gases and efficient filtration strategies
- Defect reduction, continuous yield improvements and keeping cost per transistor low is the key to the industry's success
- Contaminations and Excursions will have significant financial impact, Therefore innovative Early Detection methods and improving the metrology gap are needed.
- Several promising technologies are emerging to extend moore's law into sub 10nm technology nodes:
  - Preventing pattern collapse
  - Integrating nanowire transistor channel for GAA, fin thinning, and III-V elements devices
  - Reducing effects of short noise
  - Development of environmentally friendly solutions

SAMSUNG
AUSTIN SEMICONDUCTOR

Acknowledgments

- Colleagues at SAS
- To All SAS Suppliers: Thank You for Your Valued Partnership
- To the Linx Consulting Group, Thank You for the Opportunity to Present

SAMSUNG
AUSTIN SEMICONDUCTOR

Anton Paar

The Zeta Potential of Conductive Material Surfaces: Challenges and Solutions

Thomas Luxbacher

Anton Paar GmbH

Anton-Paar-Strasse 20, 8054 Graz, Austria

+43 316 257 2730, thomas.luxbacher@anton-paar.com

Anton Paar

Contents

- Why zeta potential?
- Electrophoretic mobility
- Streaming potential
- Conductive coating

SPCC 2020 virtual conference | September 2, 2020

Anton Paar

Why zeta potential?

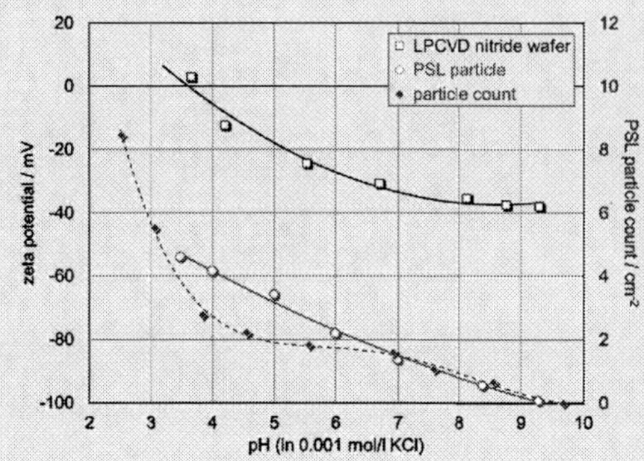

D. Jan & S. Raghavan, J. Electrochem. Soc. 141 (1994) 2465

SPCC 2020 virtual conference | September 2, 2020

Anton Paar

Why zeta potential?

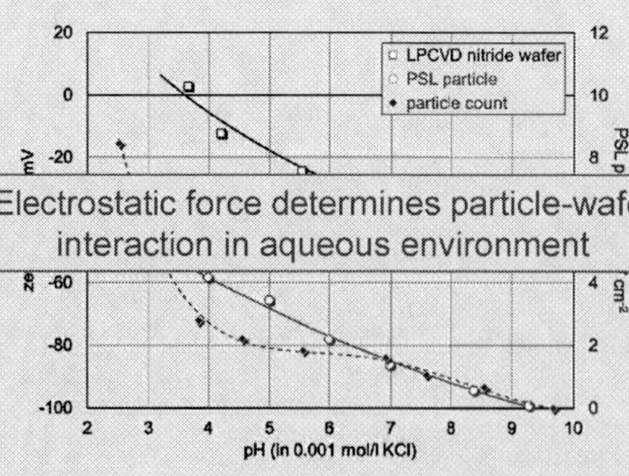

Electrostatic force determines particle-wafer interaction in aqueous environment

D. Jan & S. Raghavan, J. Electrochem. Soc. 141 (1994) 2465

SPCC 2020 virtual conference | September 2, 2020

Anton Paar

Electrophoretic mobility

- Electric field induces a collective motion of particles
- Highly charged particles move faster than less charged particles

$$\zeta = \underbrace{\frac{v}{E}}_{\mu_e} \times \frac{\eta}{\varepsilon_r \times \varepsilon_0}$$

SPCC 2020 virtual conference | September 2, 2020

Anton Paar

CMP slurry

Zeta potential of slurry particles

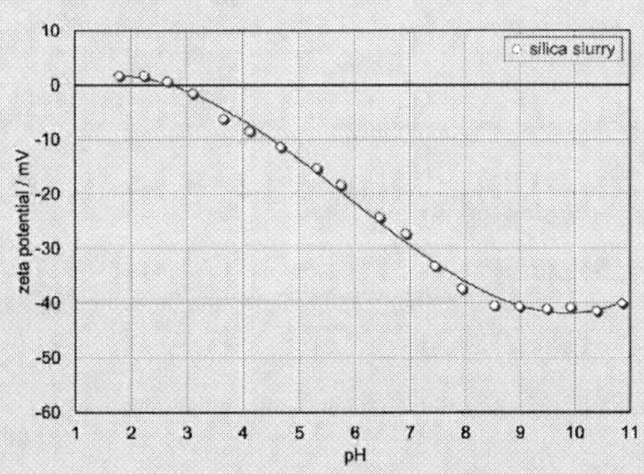

SPCC 2020 virtual conference | September 2, 2020

Anton Paar

CMP slurry
Zeta potential of slurry particles

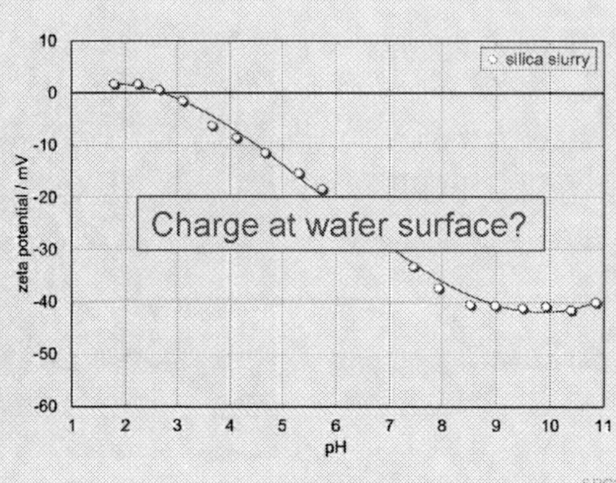

SPCC 2020 virtual conference | September 2, 2020

Anton Paar

Principle of surface zeta potential
Streaming potential | Tangential mode

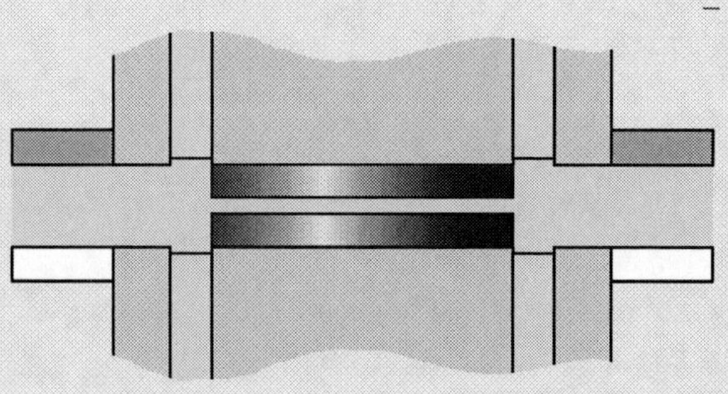

- Solid sample arranged to create a capillary channel

SPCC 2020 virtual conference | September 2, 2020

Anton Paar

Principle of surface zeta potential

Streaming potential | Tangential mode

– Solid sample arranged to create a capillary channel

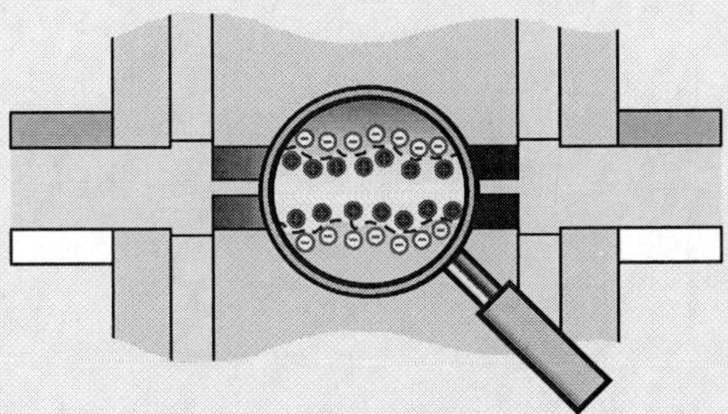

SPCC 2020 virtual conference | September 2, 2020

Anton Paar

Principle of surface zeta potential

Streaming potential | Tangential mode

– Solid sample arranged to create a capillary channel

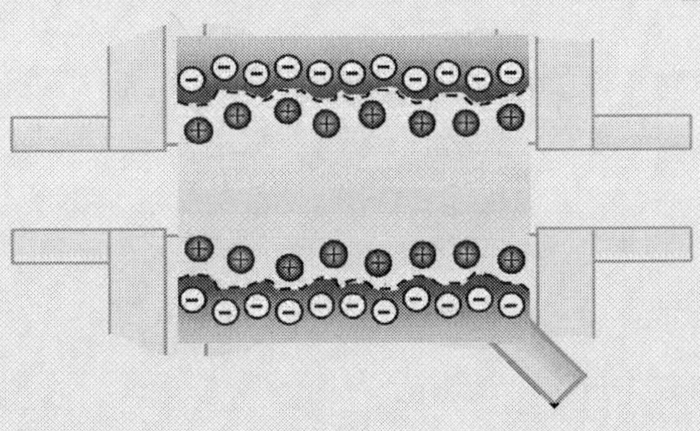

SPCC 2020 virtual conference | September 2, 2020

Anton Paar

Principle of surface zeta potential
Streaming potential

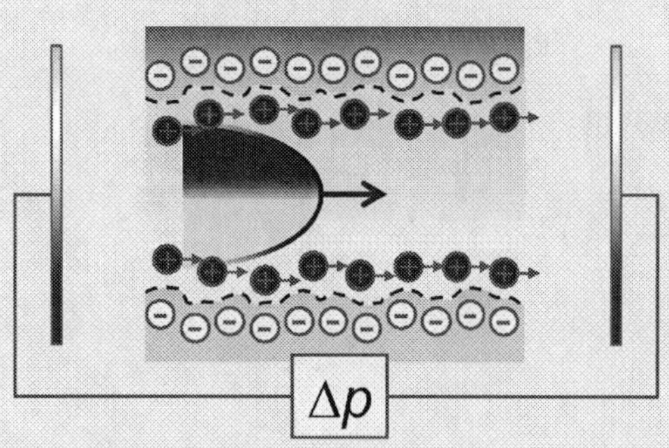

- Solid sample arranged to create a capillary channel
- Pressure gradient provokes liquid flow
- Charge distribution at solid | liquid interface distorted

SPCC 2020 virtual conference | September 2, 2020

Anton Paar

Principle of surface zeta potential
Streaming potential

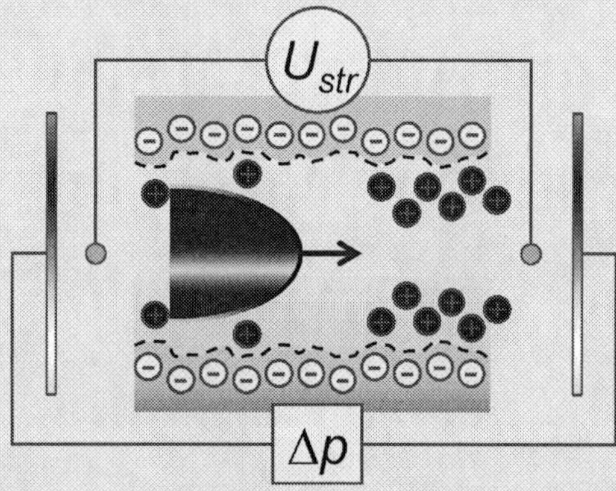

- Solid sample arranged to create a capillary channel
- Pressure gradient provokes liquid flow
- Charge distribution at solid | liquid interface distorted
- Streaming potential (d.c. voltage) generated along the capillary

SPCC 2020 virtual conference | September 2, 2020

Anton Paar

Principle of surface zeta potential

Streaming current

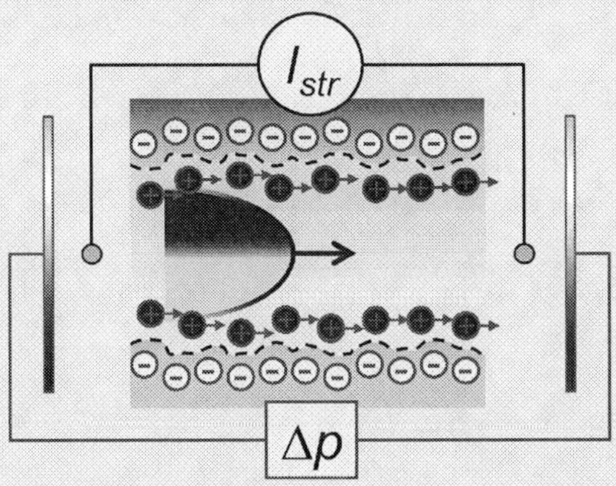

- Solid sample arranged to create a capillary channel
- Pressure gradient provokes liquid flow
- Charge distribution at solid | liquid interface distorted
- Streaming current (d.c. current) generated along the flow channel

Anton Paar

Principle of surface zeta potential

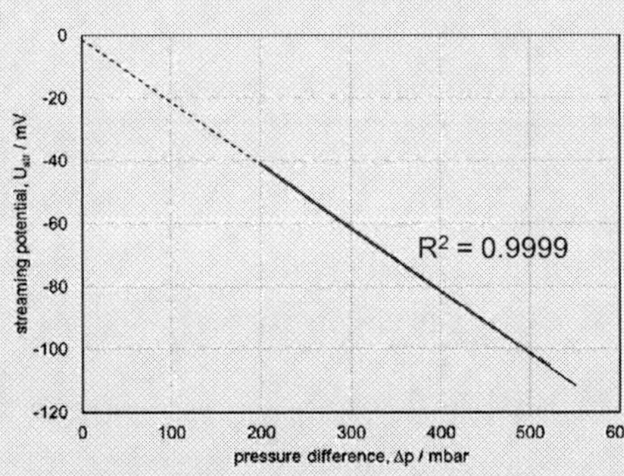

$$\zeta = \frac{dU_{str}}{d\Delta p} \times \frac{\eta}{\varepsilon_r \times \varepsilon_0} \times \kappa_B$$

$$\zeta = \frac{dI_{str}}{d\Delta p} \times \frac{\eta}{\varepsilon_r \times \varepsilon_0} \times \frac{L}{A}$$

U_{str} .... streaming potential
I_{str} .... streaming current
Δp .... pressure difference
η .... viscosity
$\varepsilon_r \times \varepsilon_0$ .... dielectric permittivity
κ_B .... electrolyte conductivity
L/A .... cell constant

Anton Paar

CMP slurry

Zeta potential of wafer surface

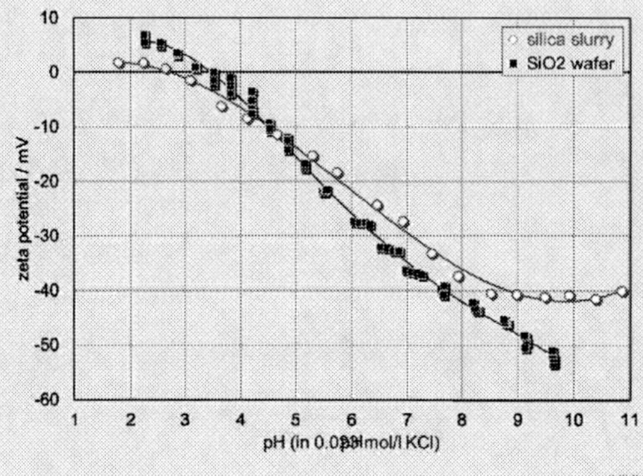

Anton Paar

Streaming potential

Surface modification of top coating

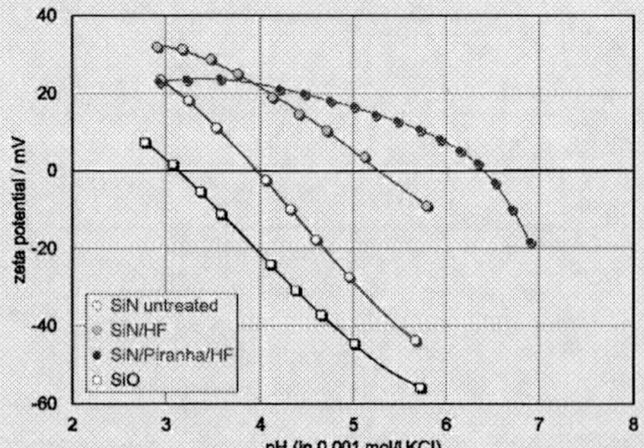

- SiN
  Si wafer with Si_3N_4 top layer
- SiO
  Si wafer with SiO_2 top layer
- SiN/HF
  Si_3N_4 after HF dip
  removal of silicon oxide
- SiN/Piranha/HF
  Si_3N_4 after Piranha etch and HF dip
  removal of organic contaminant

Anton Paar

Conductive material surface

Non-linear pressure ramp

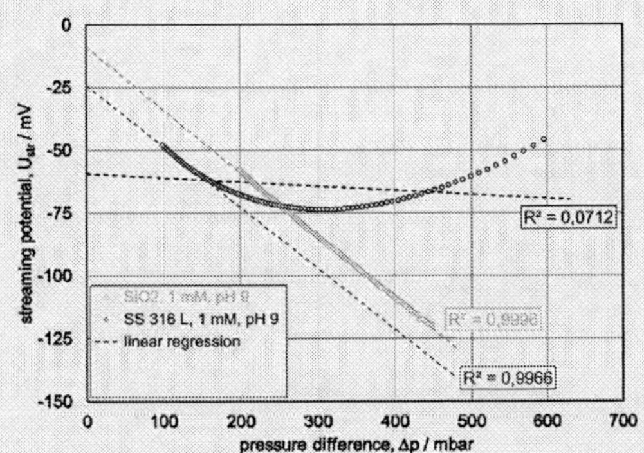

$\zeta = -3.45$ mV

$\zeta = -51.0$ mV

Anton Paar

Conductive material surface

Electric interference of material with streaming potential

- Conductive material behaves like electrode
- Charging of material surface by streaming potential (d.c. voltage)

Prediction

- Correct $dU_{str}/d\Delta p$ at zero flow
- Effect decreases with increasing ionic strength

Conductive material surface

Non-linear pressure ramp

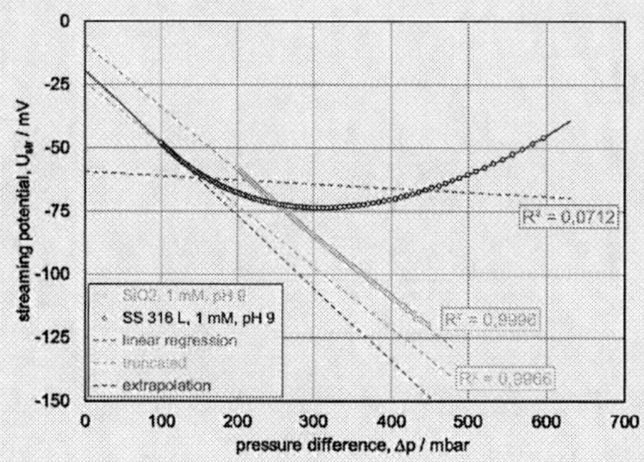

ζ = – 3.45 mV

ζ = – 51.0 mV

ζ = – 60.2 mV

Conductive material surface

Non-linear pressure ramp

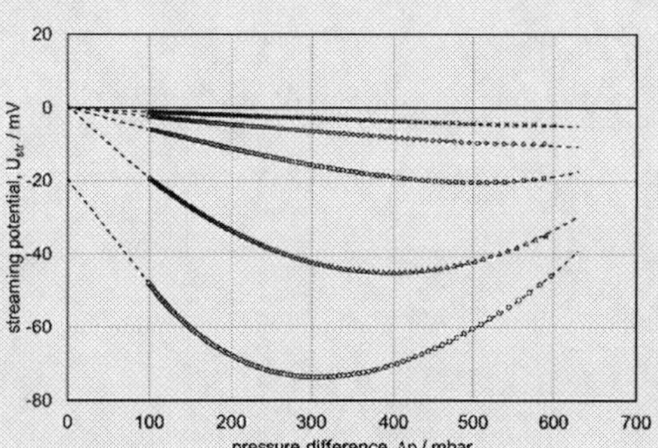

Ionic strength	R^2_{linear}
20 mM	0.996
10 mM	0.986
5 mM	0.934
2 mM	0.663
1 mM	0.071

Conductive material surface

Reliable zeta potential results

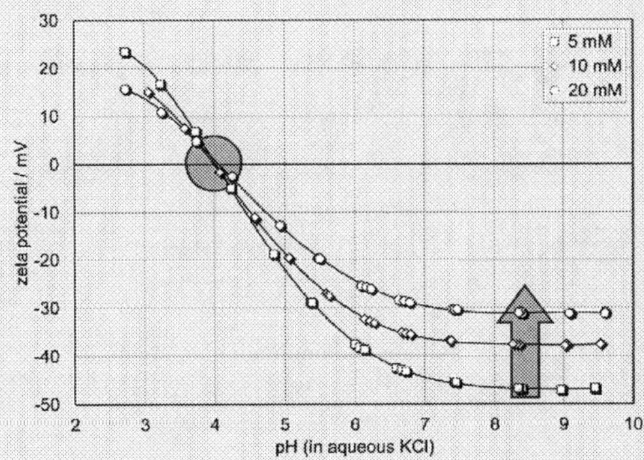

Thank you for your attention!

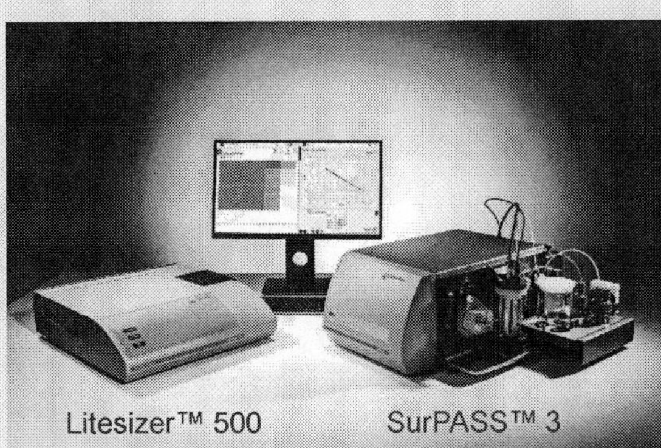

SELECTIVE Ge REMOVAL FROM SiGe SURFACES FOR Si-CAP-FREE SiGe PASSIVATION

KURT WOSTYN,, HIROAKI ARIMURA, ANDRIY HIKAVYY, DIRK RONDAS, ADRIAN VAISMAN CHASIN, THIERRY CONARD, LARS-ÅKE RAGNARSSON, NAOTO HORIGUCHI

PUBLIC

Si-CAP-FREE SiGe PASSIVATION

SiGe FIN OR GAA PASSIVATION IN RMG

Ge-free oxide IL **and** interface enables SiGe passivation
= similar to Si cap without depositing Si-cap.

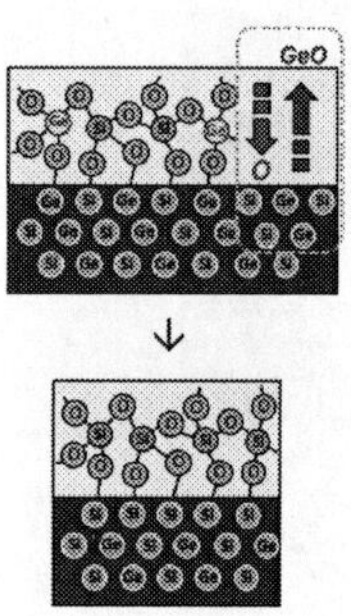

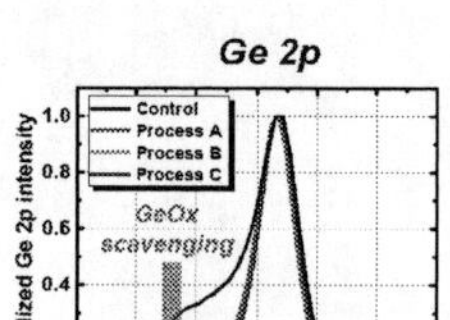

Process A, B and C:
- Removes GeO
- Variable efficiency in GeO_x removal

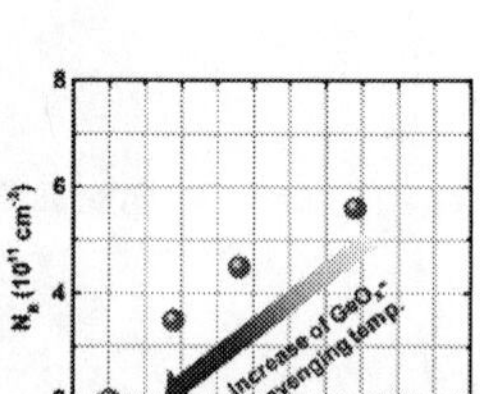

GeO_x scavenging reduces D_{IT}

[Lee et al. VLSI 2016 and Siddiqui et al. MRS Spring 2016]

Si-CAP-FREE SiGe PASSIVATION

IMPACT OF SURFACE PREPARATION PRIOR TO IL GROWTH

Ge conc at SiGe surface

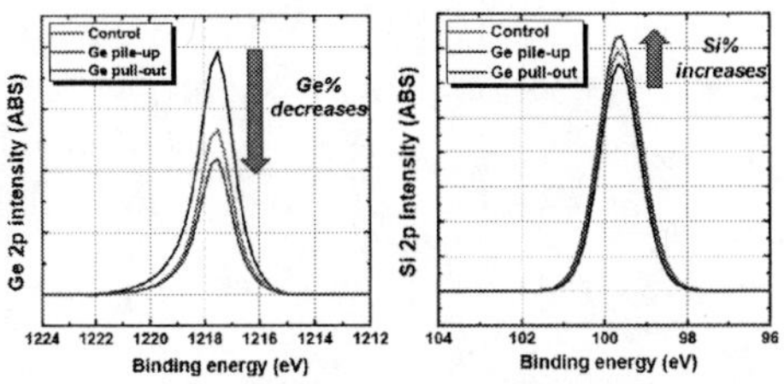

- Ge poor surface == Ge pull-out
- Ge rich surface == Ge pile-up

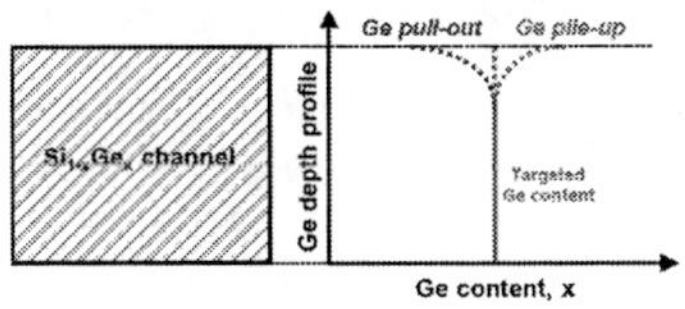

Electrical impact

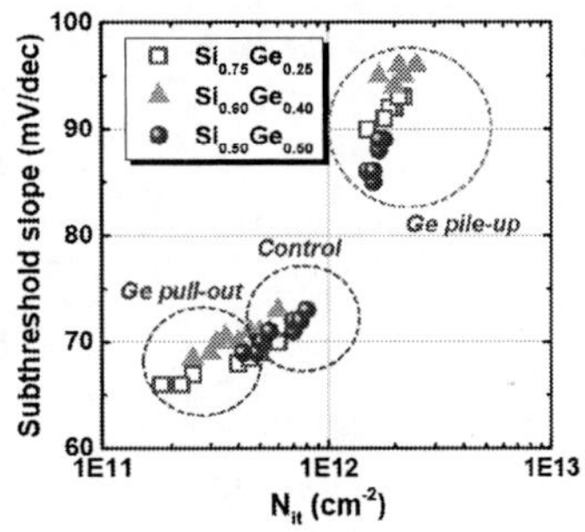

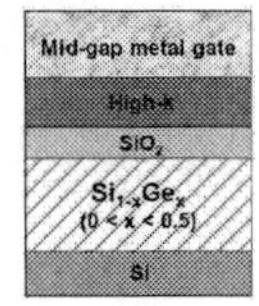

(b)

Ge poor surface, i.e. Ge pull-out, results in best electrical performance. Unfortunately, no process information was shared.

imec

[Lee et al. IEDM Tech Dig (2016) p. 766]

OUTLINE

- INTRODUCTION
- Ge PULL-OUT – SELECTIVE REMOVAL OF Ge FROM SIGE SURFACE
- IMPACT GE PULL-OUT ON Si-CAP-FREE SiGe PASSIVATION
- CONCLUSION

imec

SELECTIVE Ge REMOVAL FROM SiGe SURFACE
INSPIRED BY Si GAA SELECTIVE ETCH

HCl (G) SELECTIVE ETCH

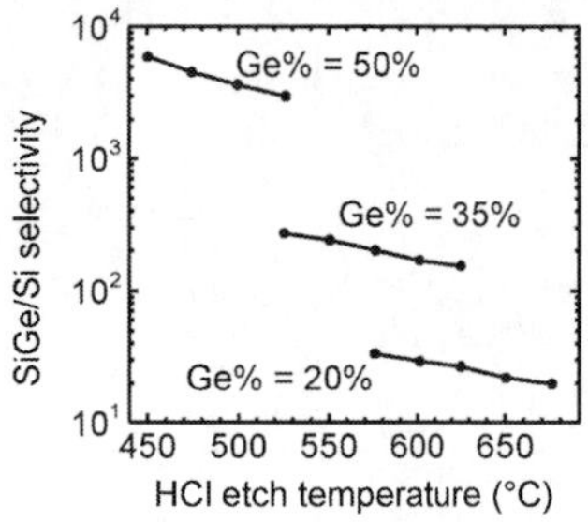

[Destefanis et al., Semicond. Sci. Technol. 23, (2008) 105019]

Si GAA BY HCl (G)

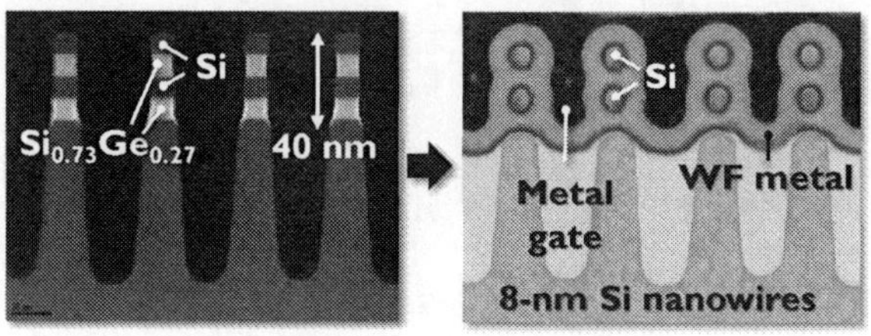

[H. Mertens et al. VLSI Tech. Dig. (2016) p. 158]

- SiGe vs Si selective etch in HCl (g) enables Si GAA
- Selectivity attributed to enhanced $GeCl_2$ desorption compared to $SiCl_2$ in HCl (g) ambient.

→ Possible to use HCl (g) for surface preparation?

imec

Ge PULL OUT – SELECTIVE Ge REMOVAL
NATIVE OXIDE COMPOSITION AFTER SURFACE TREATMENT

SiGe epi → HF (aq) → HCl (g) → Air exposure → Native oxide → Ge% from Ge3d and Si2p peaks

ANGLE-INTEGRATED XPS

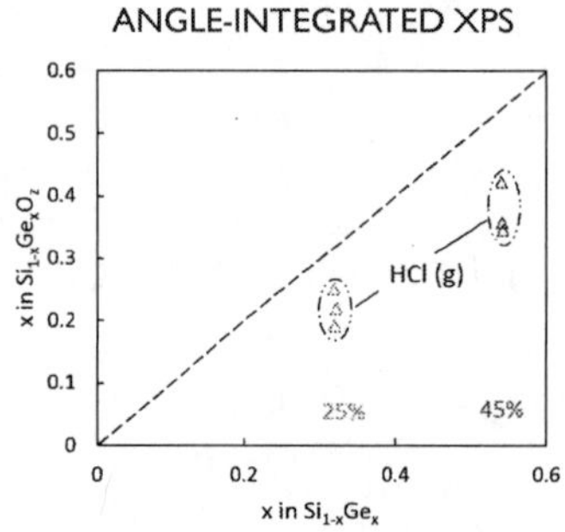

- SiGe surface becomes Ge poor in HCl (g) ambient.

imec

Ge PULL OUT – SELECTIVE Ge REMOVAL
NATIVE OXIDE COMPOSITION AFTER SURFACE TREATMENT

SiGe epi
|
HF (aq)
|
ASM Previum™
|
HCl or Cl_2 (g)
|
Air exposure
|
Native oxide
↓
Ge% from Ge3d and Si2p peaks

ANGLE-INTEGRATED XPS

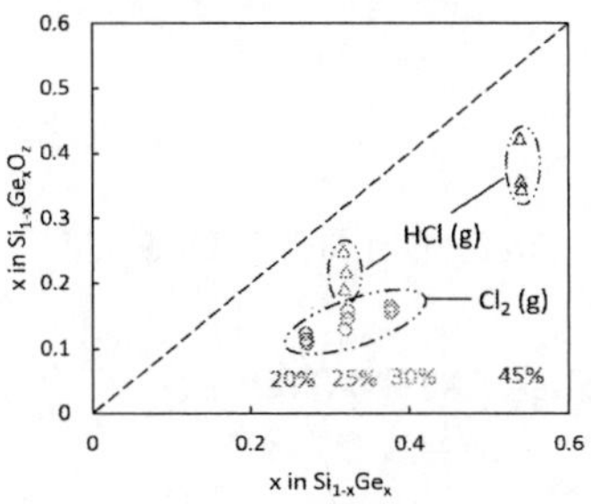

IMPACT OF Cl_2 (g) TREATMENT TIME

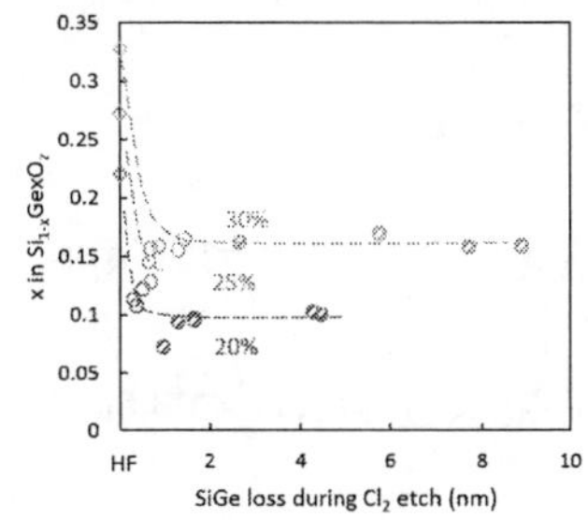

- SiGe surface becomes Ge poor in HCl (g) and Cl_2 (g) ambient.
- Tuning process conditions allows for SiGe loss reduction while maintaining selective Ge removal → surface treatment.

imec

Ge PULL OUT – SELECTIVE Ge REMOVAL
NATIVE OXIDE COMPOSITION AFTER SURFACE TREATMENT

SiGe epi
|
HF (aq)
|
ASM Previum™
|
HCl or Cl_2 (g)
|
Air exposure
|
Native oxide
↓
Ge% from Ge3d and Si2p peaks

ANGLE-INTEGRATED XPS

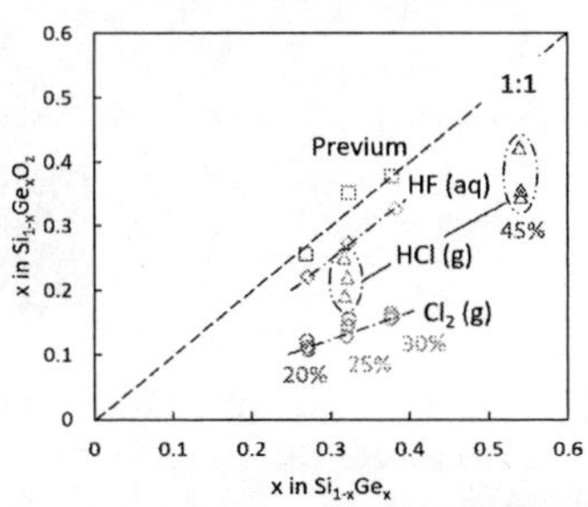

AQUEOUS HF SOLUTIONS

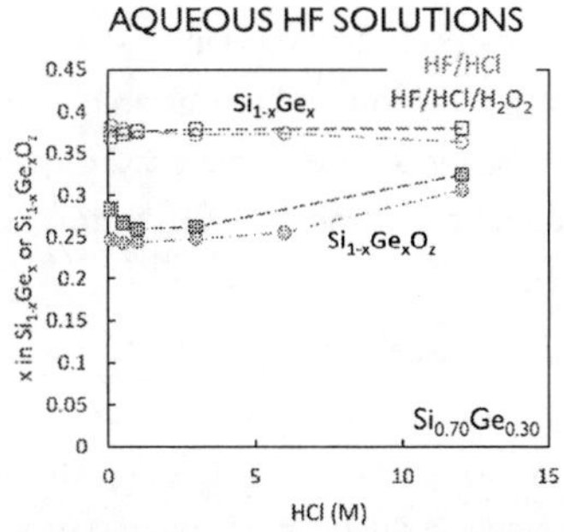

- SiGe surface becomes Ge poor in HCl (g) and Cl_2 (g) ambient.
- Tuning process conditions allows for SiGe loss reduction while maintaining selective Ge removal → surface treatment.
- Ge pull-out also observed in aqueous HF solutions. Ge pull-out efficiency depends on [HCl] and [H_2O_2], however never as efficient as Cl_2 (g).

imec

OUTLINE

- INTRODUCTION
- Ge PULL-OUT – SELECTIVE REMOVAL OF Ge FROM SIGE SURFACE
- IMPACT GE PULL-OUT ON Si-CAP-FREE SiGe PASSIVATION
- CONCLUSION

imec

SiGe PASSIVATION – HOW TO? CHALLENGES.
Ge-OXIDE-FREE INTERLAYER

- Avoid Ge-oxide formation
  - Si cap – reference passivation method
  - Thermal interlayer – pure SiO_2 formed by Ge condensation
  - Ge pull-out prior to IL formation
- Ge-oxide scavenging
  - High solubility of GeO_2 in aqueous solutions
  - Scavenging prior to high-k deposition
  - Scavenging during post-metallization anneal (PMA)

  } Imec POR

Ideal SiGe passivation

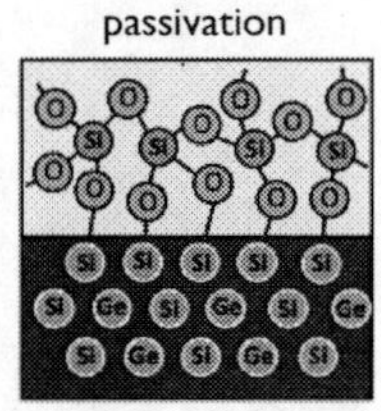

[Lee et al. VLSI 2016]

Oxide and interface composition:
- Pure SiO_2
- Interface contains only Si – O bonds
- No Ge gradient below interface

imec

10

GeO SCAVENGING

NH_3 ANNEAL + H_2 ANNEAL ON CHEMOX

[H. Arimura et al. IEDM Tech Dig (2019) p. 667 and K. Wostyn et al. presented at ISTDM/ICSI 2019]

Limited delay

- SiGe 25-30% epi
- Chem ox.
- NH_3 anneal (700°C, 5 torr, 60 s)
- H_2 anneal (775°C, 40 torr, 2 min)
- Water rinse
- HfO_2 (2 nm)
- PVD TiN/W gate
- HPA: 400C, 20 ATM, H_2

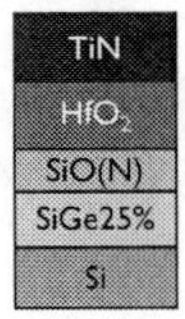

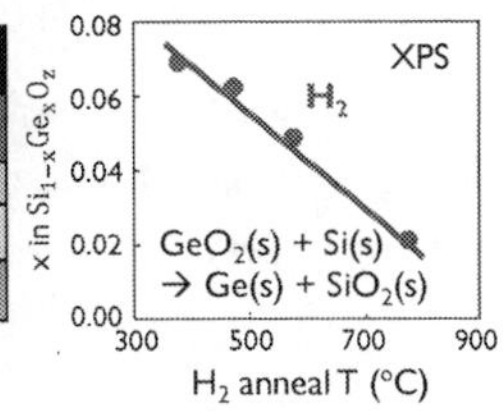

- H_2 anneal at high-T on SiO_2 (775°C, 40 torr, 2 min)
  - SiO vs GeO thermal stability difference
  - **GeO_2(s)** + Si(s) → Ge(s) + **SiO_2(s)**
- Additional NH_3 anneal on SiO_2 (700°C, 5 torr, 60 s)
  - Nitridation to suppress O / Vo diffusion and GeOx formation [J. Huang et al., APL 88, 143506 (2006).]
  - Increase in k-value (SiON)

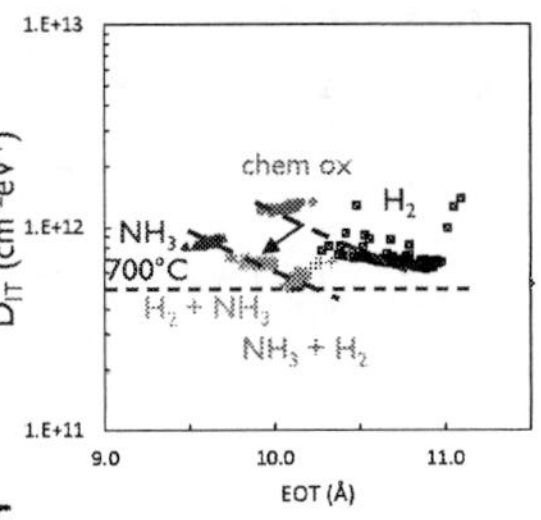

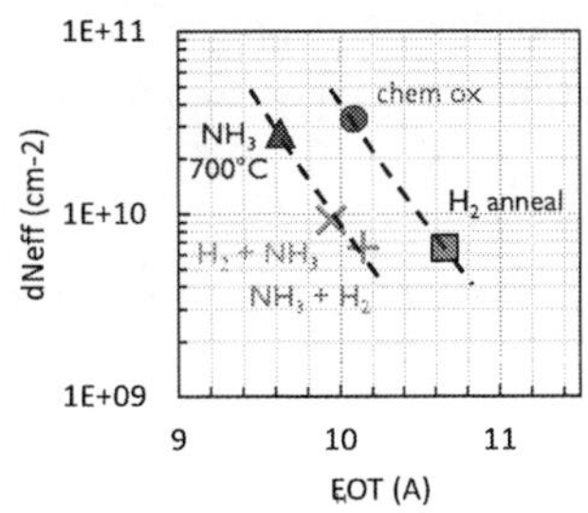

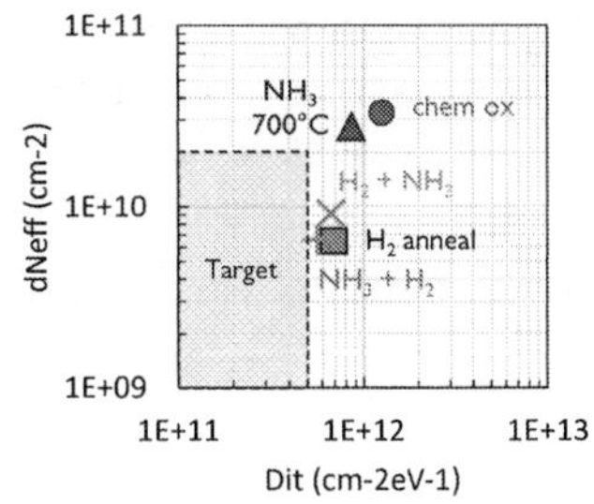

imec

IMPACT OF GE PULL-OUT ON $Si_{0.75}Ge_{0.25}$ PASSIVATION

Ge PULL-OUT IN HCl (G) AMBIENT

SiGe passivation

- SiGe 25% epi
- HF 0.7% 60s
- **Ge pull-out: HCl (g)**
- O_3 (aq) chem ox
- GeO_x scavenging
- Water rinse
- HfO_2 deposition
- TiN/W gate
- Patterning
- HPA: 400C, 20 ATM, H_2

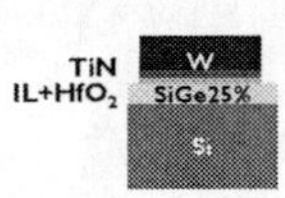

C-V hysteresis measurements at RT

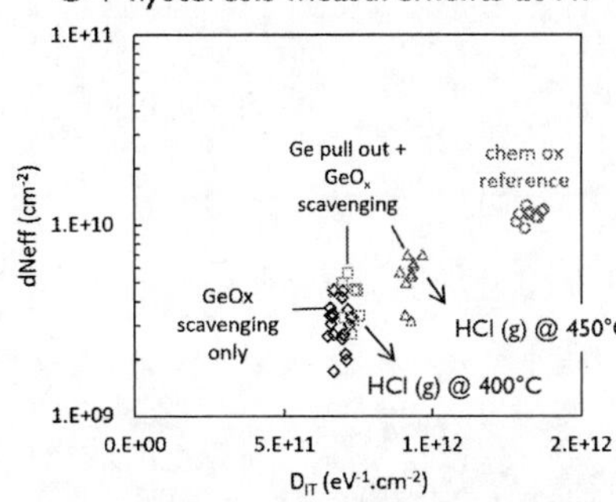

- GeO_x scavenging improves SiGe passivation.
- Combining Ge oxide scavenging with Ge pull-out using HCl (g) shows no or small detrimental impact on dNeff vs Dit.

imec

CONCLUSIONS

- Ge pull-out observed for SiGe < 50% in
  - Aqueous HF solutions, HCl (g) and Cl_2 (g) on oxide-free surface
  - Ge pull-out most efficient using Cl_2 (g)
- Si-cap-free SiGe passivation
  - Improved SiGe passivation by combining chem oxide with Ge-oxide scavenging
  - No or small detrimental impact observed when Ge pull-out in HCl (g) combined with Ge-oxide scavenging
- Future: impact of Ge pull-out on SiGe oxidation → come and see at UCPSS!

imec 13

imec

embracing a better life

imec PUBLIC

Ge PULL OUT – SELECTIVE Ge REMOVAL

NATIVE OXIDE COMPOSITION AFTER SURFACE TREATMENT

SiGe epi
|
HF (aq)
|
ASM Previum™
|
HCl or Cl_2 (g)
↓
Air exposure

Native oxide

ANGLE-INTEGRATED XPS

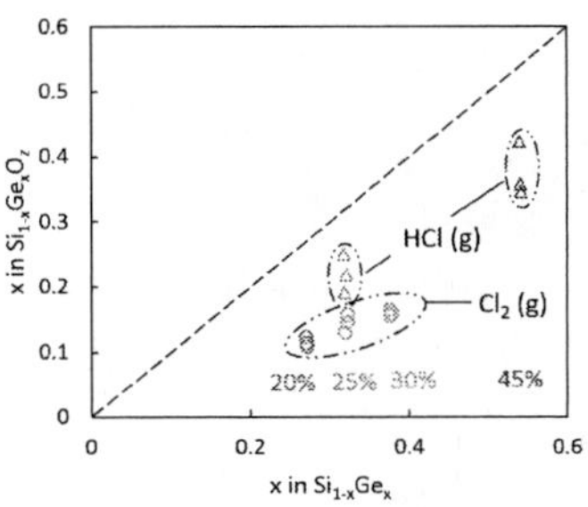

DESORPTION OF HCL (ADS) ON GE (100)

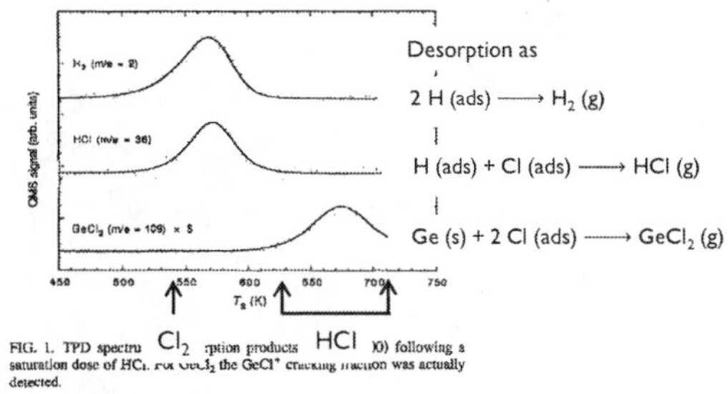

[D'Evelyn et al. J. Chem. Phys. 101 (1994) 2463]

Difference in surface Ge conc between HCl and Cl_2 possibly (partially) related to difference in T during surface preparation step.

imec

SURFACE PREPARATION OF NIOBIUM FOR SUPERCONDUCTING QUBIT APPLICATIONS

ANTOINE PACCO

ANTOINE.PACCO@IMEC.BE

ANTON POTOCNIK, JEROEN VERJAUW, MASSIMO MONGILLO, TSVETAN IVANOV, ROHITH ACHARYA, DANNY WAN, BOGDAN GOVOREANU, IULIANA P. RADU, KURT WOSTYN, EFRAIN ALTAMIRANO SANCHEZ AND FRANK HOLSTEYNS

PUBLIC

QUBIT DEVICE FOCUS @ IMEC

SEMI- & SUPER-CONDUCTING QB'S

	No issue
	Good enough
	Survivable
	Not clear, difficult
	Showstopper

Qubit Type	NMR	Ion traps/ atoms	semiconductor QD	N V in diamond*	Superconduct. circuits	Optical qubits	Gatemon (Majorana)
Coherence/ gate time	OK [illegible]	OK [illegible]	OK -E7	OK ~E7	OK 5E7	OK ~E5	OK E5
2D array path		likely	Likely with extreme scaling	likely	yes	likely	maybe
Max qb #	14	[illegible]	2	2	21 (-2000)	?	
Fabrication	chemistry	Maybe Si fab likely*	CMOS compatible	Random Single ion	$Al/Al_2O_3/Al$ Si fab OK	[illegible] platform	CMOS compat Maybe
Device size		Atom but many lasers	transistor	Not clear	100s of um²	mm²	mm²
other	Requires very large magnets	Requires lasers 4K	25mK	Requires lasers 4K	25mK	Deterministic 1-photon source 4K	25mK

QUBIT COHERENCE

MOORE'S LAW FOR QUANTUM COHERENCE

LOW LOSS = HIGH Q-FACTORS ← **LONG COHERENCE TIMES**

→ REDUCTION OF TWO-LEVEL-SYSTEMS

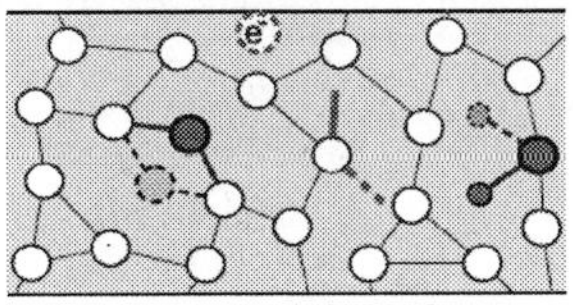

LISENFELD, SCIENTIFIC REPORTS MARCH 2016

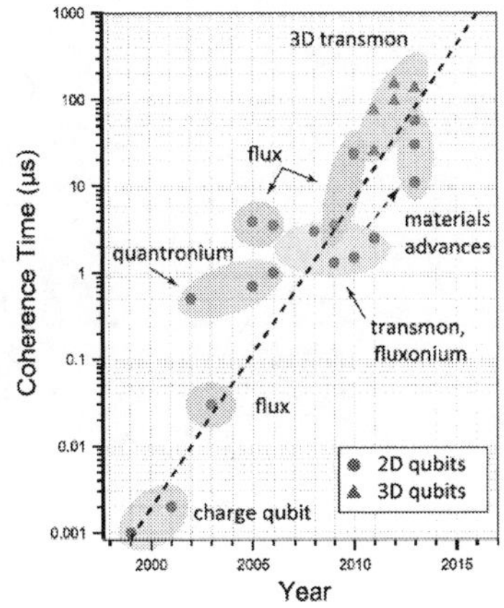

WELANDER, MRS BULLETIN VOL 38 OCTOBER 2013

imec

RESONATORS FOR DETECTION OF TLS'S

POWER DEPENDENCE OF THE INTERNAL QUALITY FACTOR Q_i

Low power ("single photon") regime is of interest to qubits

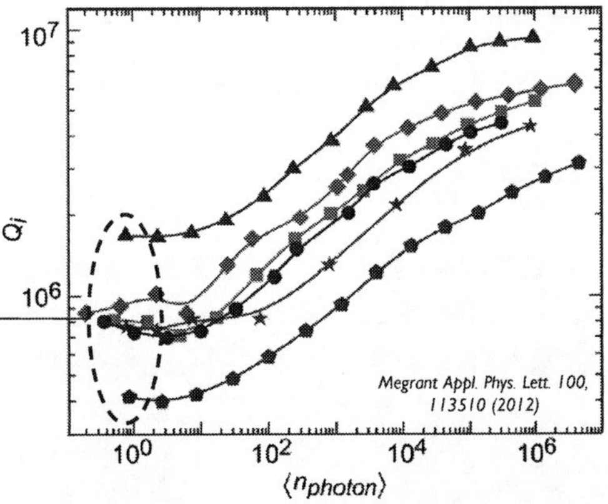

Q_i → measure for *MATERIAL* and *FABRICATION PROCESS* **QUALITY**

imec

QB FABRICATION PROCESSES AND SC METALS

IMEC 300 mm wafer FAB QUBIT PROCESS flow:

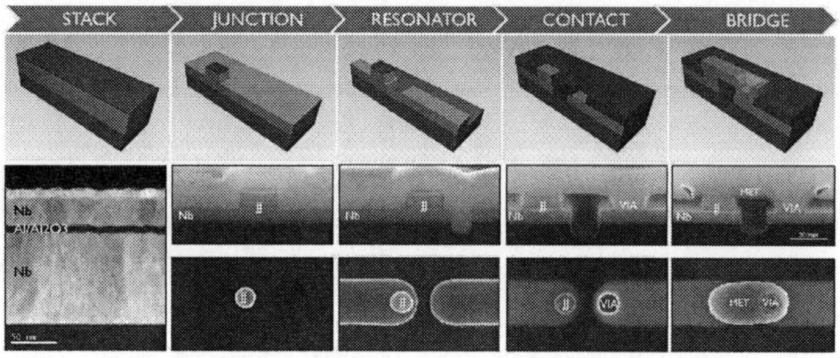

FAB PROCESSES have an influence on SURFACES & INTERFACES

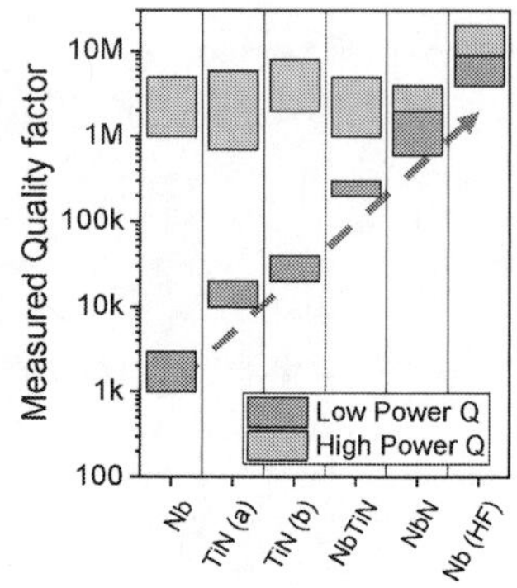

STRONG **MATERIAL &**
PROCESS DEPENDENCE

imec

NIOBIUM AS SUPERCONDUCTING METAL

PRO Promising candidate for **VLSI** and high $\mathbf{T_c}$ at **9.2 K.**
CONTRA forms a **native oxide film once exposed to air** → detrimental TLS's

THIS STUDY

EXPERIMENTAL METHOD:

RESONATOR FAB PROCESS → SURFACE TREATMENT → AGING TIME, REOXIDATION → Q FACTOR & XPS MEASUREMENTS

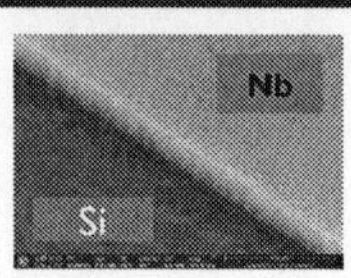

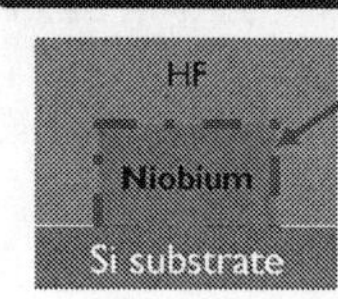

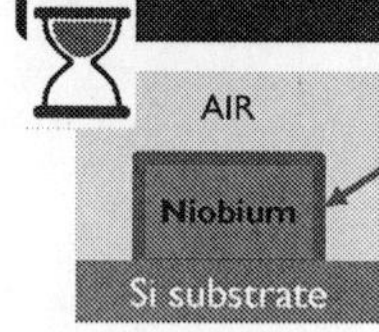

imec

CHEMICAL PROPERTIES OF NIOBIUM

- Noble? **NO**
- Resistant to corrosion? **YES**

→ **EASY** to clean niobium

→ **DIFFICULT** to have niobium in its metallic form

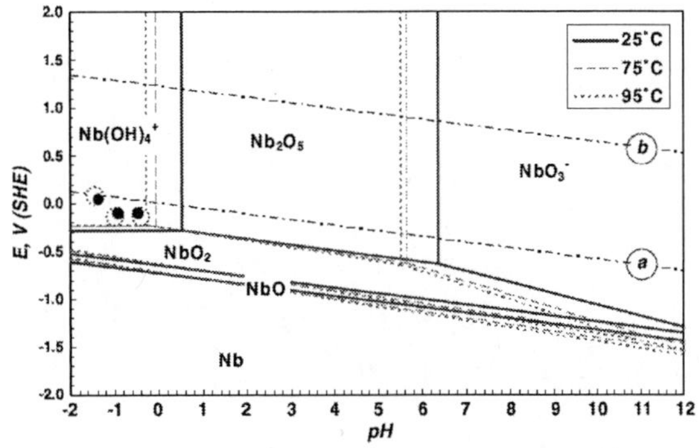

Asselin, E., Ahmed, T.M., and Alfantazi, A. (2007). "Corrosion of Niobium in Sulfuric and Hydrochloric Acid Solutions at 75 and 95 deg. C." *Corrosion Science*, 49(2), pp 694-710.

imec

7

CLEANING & ETCHING SOLUTIONS

ETCH RATE AND SURFACE COMPOSITION AFTER ETCH

Solution	Dip time (min)	Film loss (nm)	Etch Rate (nm/min)
HF	10	14.5	1.4
HF:HCl	10	10.4	1.0
BHF	10	2.3	0.2
$HF:HNO_3$	3.5	61.8	17.7

HF-based treatments etch Nb and decrease the Nb_2O_5 oxide.

$Nb_2O_5 + 10F^- + 10H^+ = 2NbF_5 + 5H_2O$

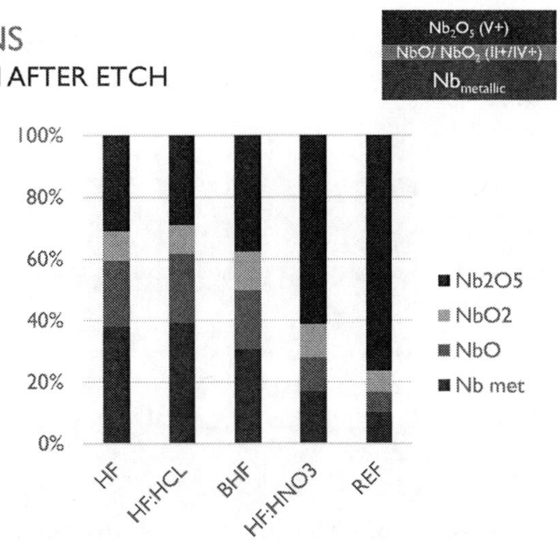

imec

8

RE-OXIDATION OF NIOBIUM (LONG TIMES)

SURFACE COMPOSITION MONITORED BY XPS (AFTER HF-TREATMENT)

Nb_2O_5 (V+)
NbO/ NbO_2 (II+/IV+)
$Nb_{metallic}$

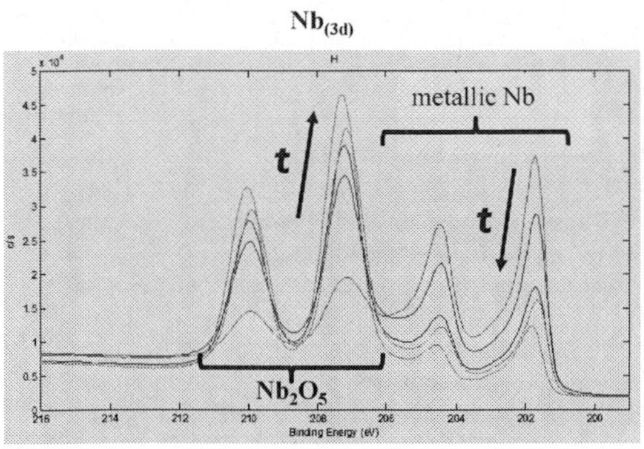

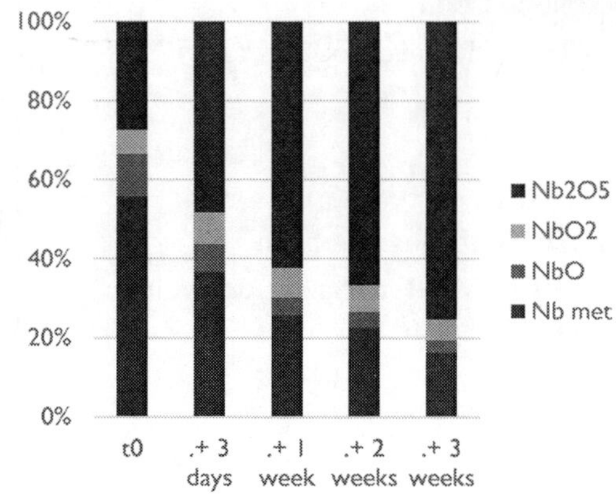

Nb gradually oxidizes to Nb_2O_5 oxide.
Sub-Ox decrease marginally a.f.o. time.

imec

RE-OXIDATION OF NIOBIUM (SHORT TIMES)

SURFACE COMPOSITION MONITORED BY XPS (AFTER HF-TREATMENT)

Nb_2O_5 (V+)
NbO/ NbO_2 (II+/IV+)
$Nb_{metallic}$

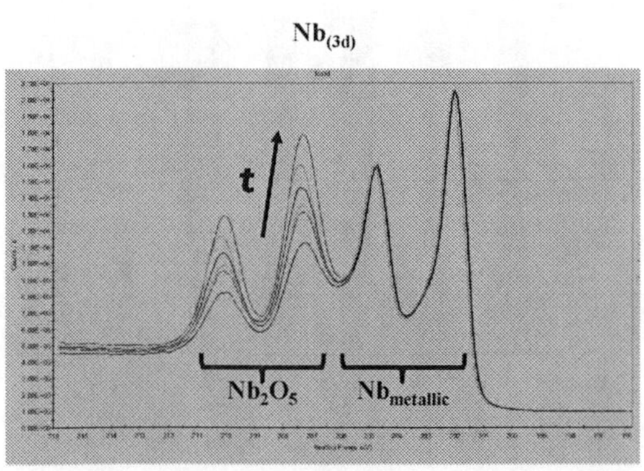

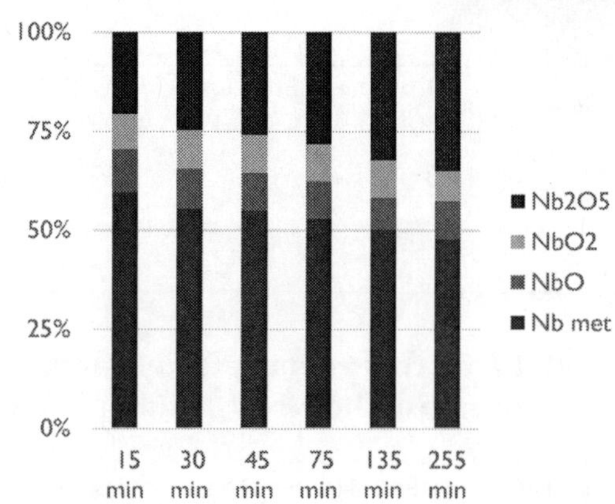

3 oxide types already present after the shortest ageing time !

imec

10

INTERNAL QUALITY FACTOR MEASUREMENTS

LE resonator

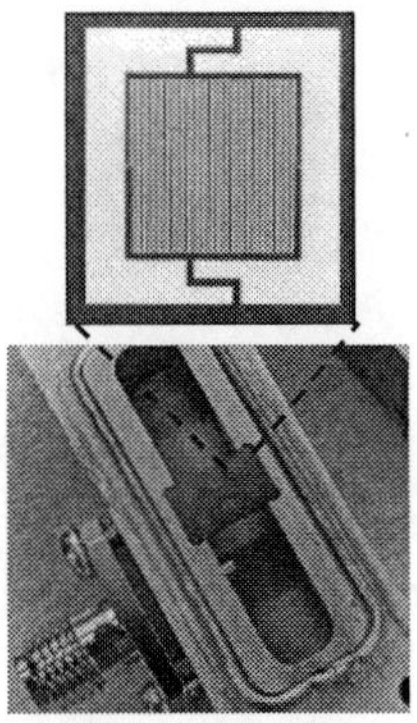

LE resonators in a 3D cavity

one of the LE resonances

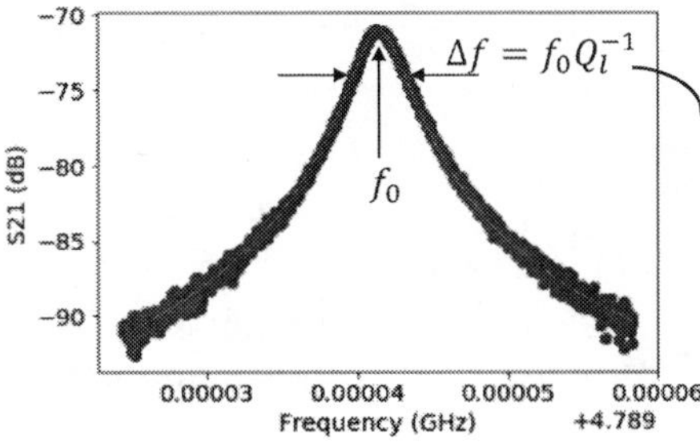

Keep small by design

$$\frac{1}{Q_l} = \frac{1}{Q_{\text{ext}}} + \frac{1}{\boldsymbol{Q}_{\textbf{int}}}$$

Measured

Material dependent term of interest

$$Q_l \sim \boldsymbol{Q}_{\textbf{int}}$$

imec

11

DEPENDENCE OF THE Q FACTOR ON RE-OXIDATION TIME

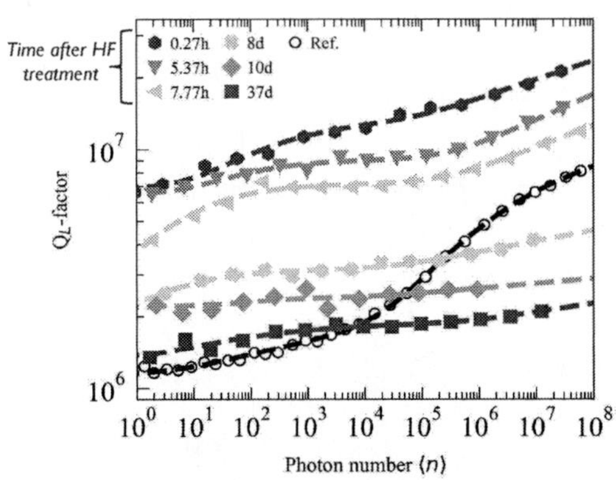

Verjauw J. et al. 300mm fab process for ultra high-quality superconducting resonators. IEDM paper, submitted.

- Q-factors increase after an HF treatment of the niobium resonators.
  - quality factors up to **7 million** in the single photon regime.
    = highest reported for niobium !
- When niobium re-oxidation takes place, the Q-factors reduce again over time.

imec

12

DEPENDENCE OF THE Q FACTOR ON Nb_2O_5 THICKNESS

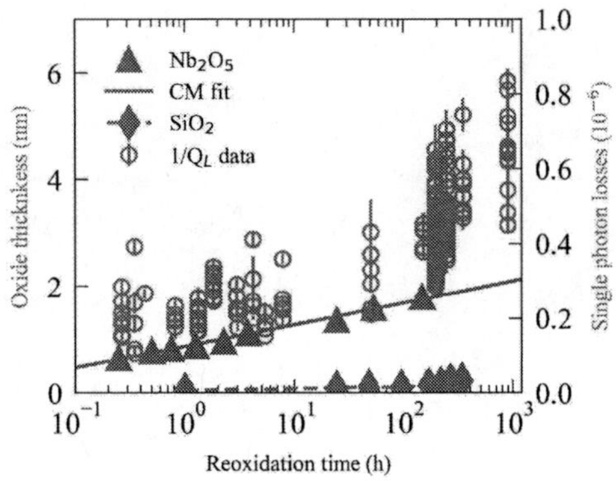

Verjauw J. et al. 300mm fab process for ultra high-quality superconducting resonators. IEDM paper, submitted.

- The lower limit of $1/Q_L$ scales with the niobium oxide thickness.

  → Reduction of Nb_2O_5 thickness improves Q factor.

- Nb_2O_5 is the main source of loss in the single photon limit within the first ~200h.

imec

13

SUMMARY

- An HF treatment partially removes the niobium(V) oxide.
- A decrease in Nb_2O_5 thickness has a huge effect on the quality factor of the niobium resonators:

  Very high quality $Q_{s.ph}$ ~ 7M were measured for the lowest aging (re-oxidation) times.

- Losses increase again with Nb_2O_5 regrowth in time.

imec

14

ACKNOWLEDGEMENTS

- Surface and Interface Processing Group:
  - Jens Rip, Kurt Wostyn, Efrain Altamirano Sanchez, Frank Holsteyns.
- Superconducting Qubit Integration Team:
  - Jeroen Verjauw, Anton Potocnik, Danny Wan, Massimo Mongillo, Tsvetan Ivanov, Laurent Souriau, Rohith Acharya, Bogdan Govoreanu, Iuliana Radu, Marc Heyns
- Materials & Component Analysis Team
  - Anja Vanleenhove, Thierry Conard, Ilse Hoflijk, Inge Vaesen

imec 15

imec

embracing a better life

imec PUBLIC

CMOS device architecture evolution and its impact on surface preparation and cleaning for N2 and beyond

Naoto Horiguchi

Imec, Leuven, Belgium

PUBLIC

OUTLINE

- CMOS device scaling challenges
- Device architecture evolution and its impact on surface preparation and cleaning
  - Nanosheet
  - Forksheet
  - Complementary FET (CFET)
  - Buried power rail
- Summary

imec

2

Current Scaling Challenges for Sub-6T cells

6T With Standard Power Rail

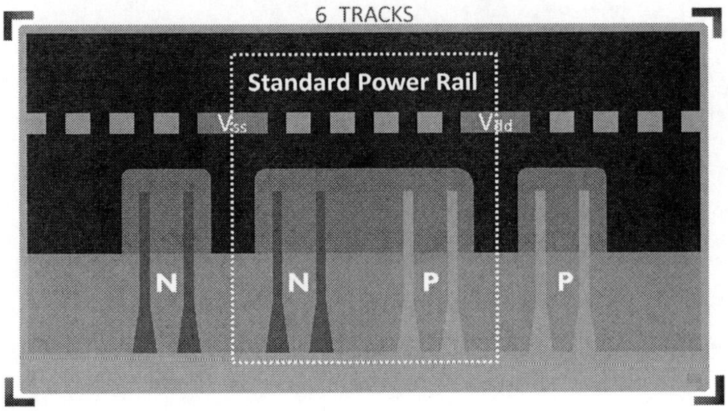

Key Challenges

- Gate & Fin pitch scaling has slowed down ➔ Compensated by **track height scaling** (= **Fin # reduction**)
- **Scaling of NP spacing**
- Need for **wide power rails** to minimize IR drop

imec

5T cells in the future

5T With Buried Power Rail (BPR)

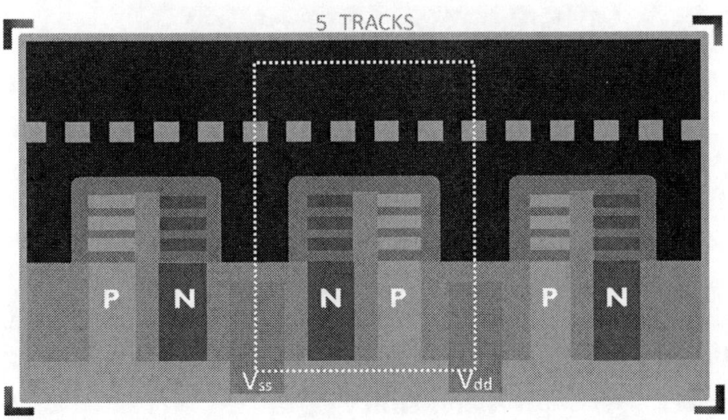

What's new

- **Single fin** architecture w/ stacked **nanosheet channel**
- Scaling of NP spacing by **Forksheet** or **CFET** (not shown).
- High aspect ratio **buried power rails** in STI/substrate

imec

Device Architecture Roadmap

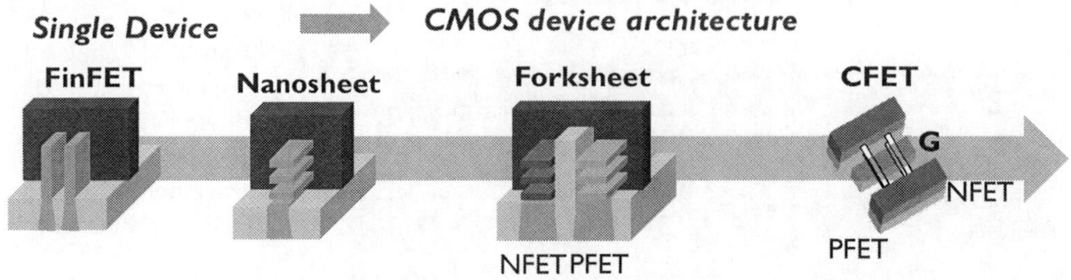

1st introduction of GAA

Evolution from 2D Nanosheet

Revolutionary 3D

imec

FINFET SCALING CHALLENGES

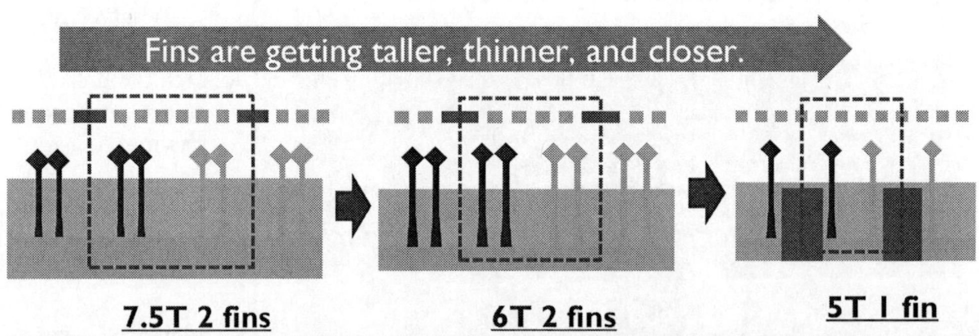

Fin de-population is required for standard cell scaling.

Drive strength ⬇ Variability ⬆

imec

EVOLUTION FROM FINFET TO NANOSHEET

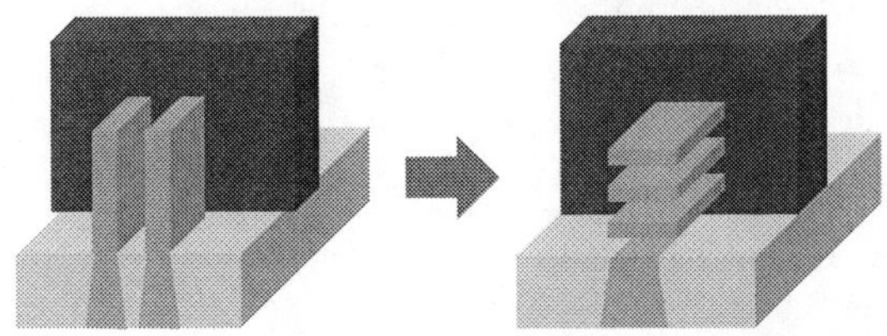

FinFET → **Nanosheet**

Nanosheet Advantages

- Maximized effective width → Drive current increase
- Improved short-channel control due to gate-all-around
- Variable device width for design flexibility

imec

Nano Sheet process flow

Blue: Nanosheet modules different from FF

- Starting material: Bulk Si wafer
- Ground plane I/I + anneal
- SiGe/Si/SiGe/Si epitaxy *(a)*
- Spacer-defined fin patterning (SADP)
- Low-temperature STI filling
- Dummy gate patterning
- RT extension I/I + Spacer formation
- Inner spacer + Embedded SD epi *(b)*
- ILD0 deposition/CMP
- Dummy poly/oxide removal
- Si channel release by SiGe etch *(c)*
- Gate dielectric (IL-SiO_2/HfO_2) dep.
- Work function (WF) metal deposition & patterning
- Fill metal (W) deposition/CMP
- Contact formation + M1 BEOL

COVENTOR

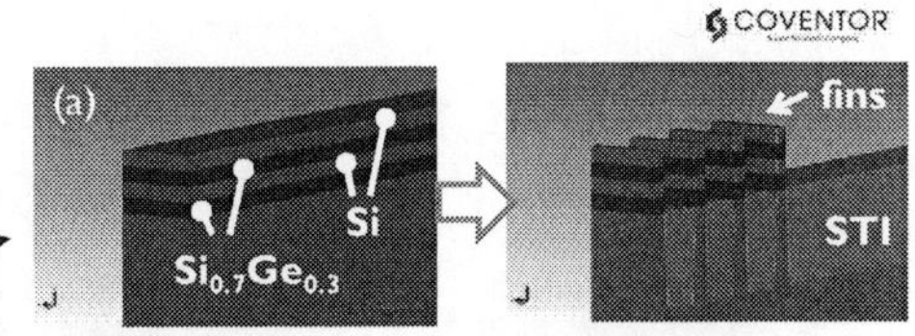

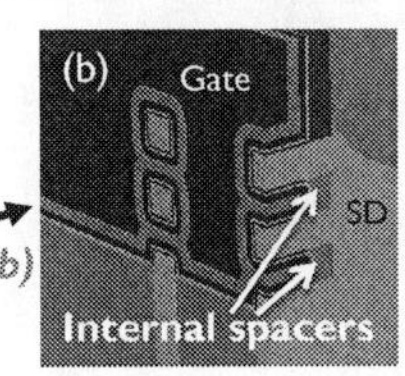

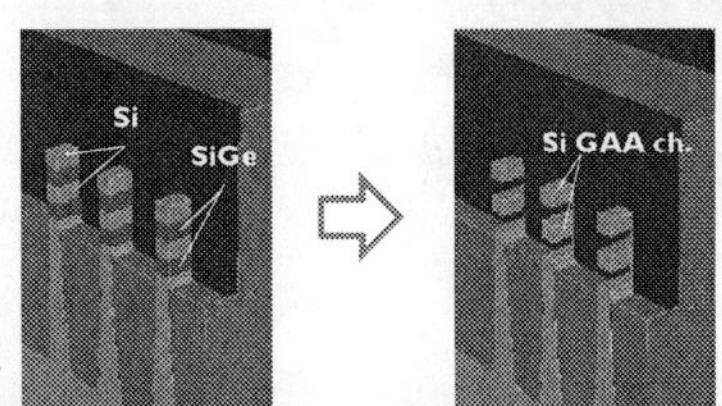

Nanosheet flow is similar as FinFETs except for Si/SiGe stack fins, inner spacer, nanosheet channel release and WFM patterning.

Y. Muraki et al SPCC 2020

Isotropic etch in nanosheet integration

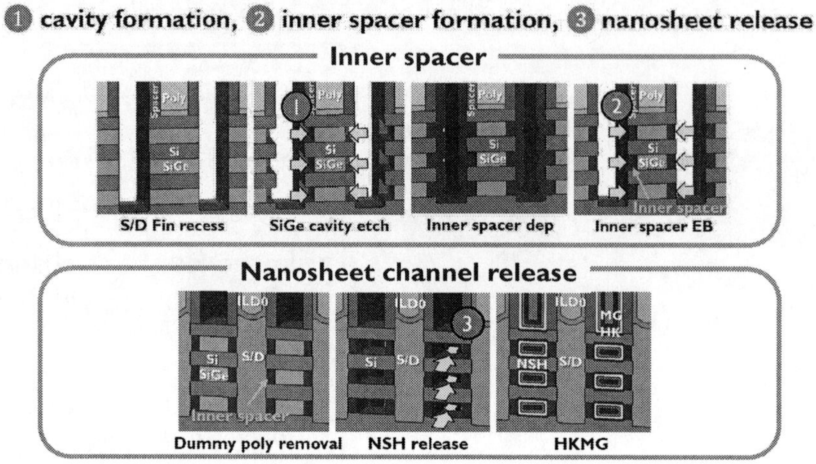

Nanosheet integration needs SiGe & dielectric isotropic selective etches.

- High etch selectivity and post etch surface clean are key.

imec

N-P SPACE SCALING BY FORKSHEET AND CFET

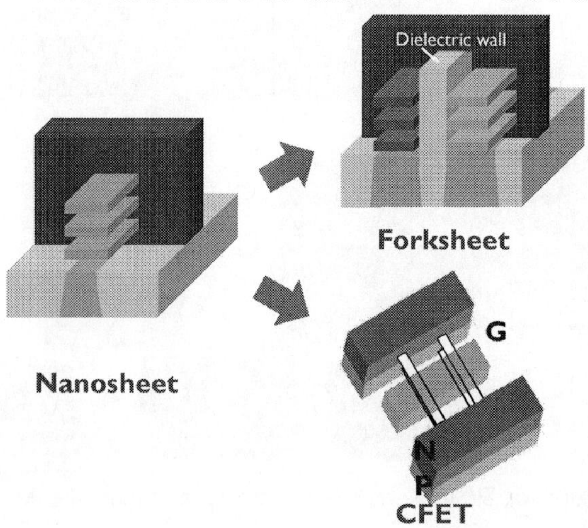

Forksheet Advantages

- N-P space can be defined by dielectric wall

CFET (Complementary FET) Advantages

- N-P space becomes vertical space due to stacked NMOS & PMOS

imec

10

Nanosheet vs Forksheet

Nanosheet + Dielectric wall = Forksheet

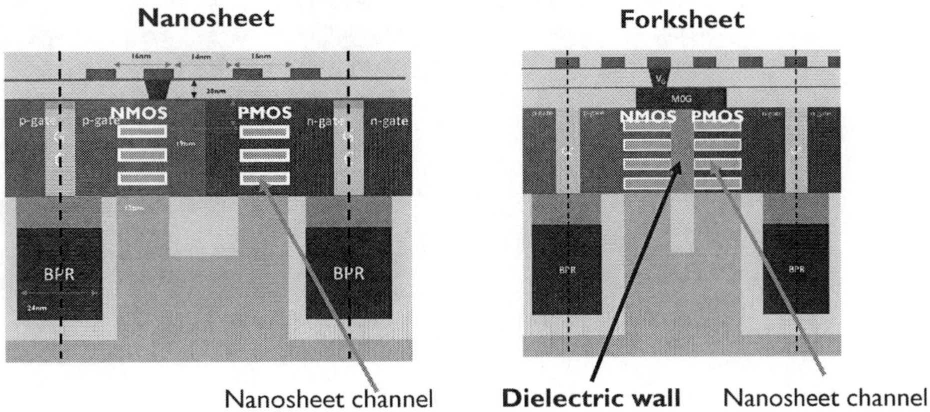

imec

Work Function Metal Patterning: Nanosheet vs Forksheet

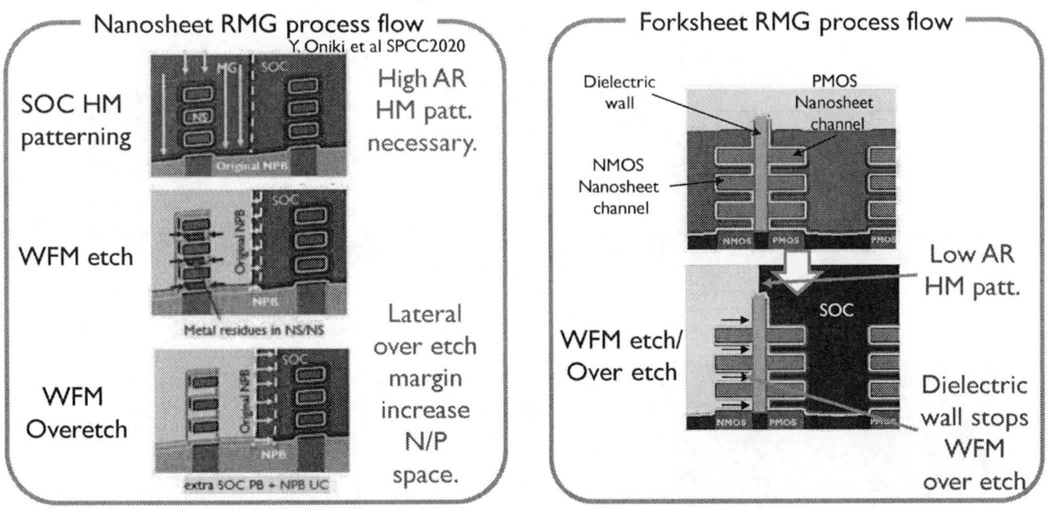

imec

Smaller N-P space is enabled by dielectric wall in forksheet.

Forksheet process flow uses backbone of nanosheet

Dielectric wall formation and its impact in subsequent processes

- Nanosheet stack patterning/STI CMP
- (BPR etch + metallization) optional
- Nanosheet stack reveal
- **FSH dielectric wall formation** → New step
- Gate patterning
- Spacer and fin recess
- **Inner spacer**
- **Source/drain epitaxy**
- ILD0 CMP
- **RMG** → modified step
- Gate recess + cut
- Gate plug
- M0A patterning + metallization
- MINT/VINT (dual damascene)

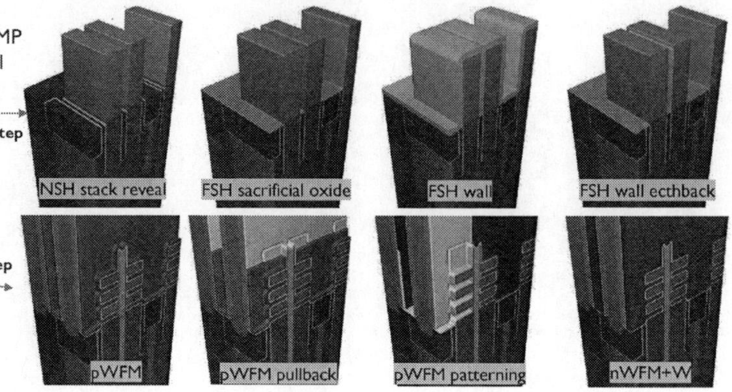

P.Weckx et al iedm 2019

Forksheet process flow is similar as nanosheet flow.

- Dielectric wall formation and its impact on subsequent module are important.

imec

13

Dielectric wall integration

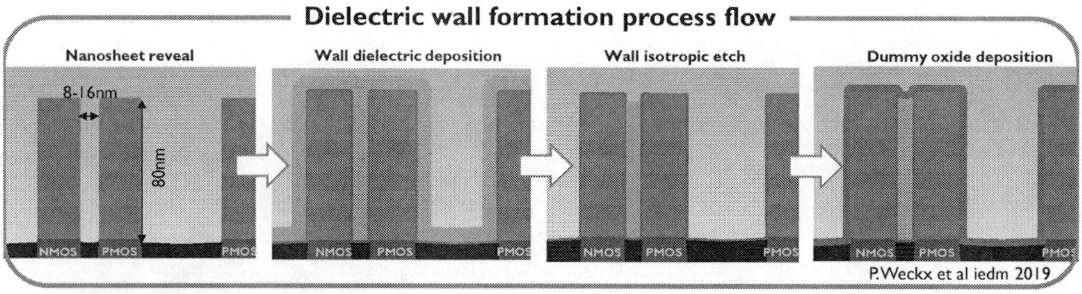

Inner spacer & CMOS SD epi HM patterning

WFM HM patterning

Forksheet key process steps

- Dielectric wall isotropic etch
- HM patterning for CMOS epi & WFM to avoid dielectric wall loss

+ Nanosheet critical steps

- Si/SiGe STI, inner spacer, channel release, WFM patterning

imec

14

Complementary FET (CFET)

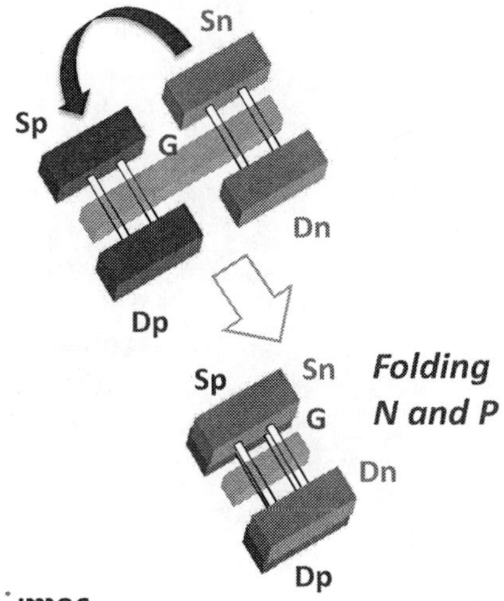

Folding N and P

- CMOS-like to single device-like foot-print
- N-P space in vertical direction has no area penalty.

Demonstration of Monolithic CFET

Bottom pFF

Top nNS

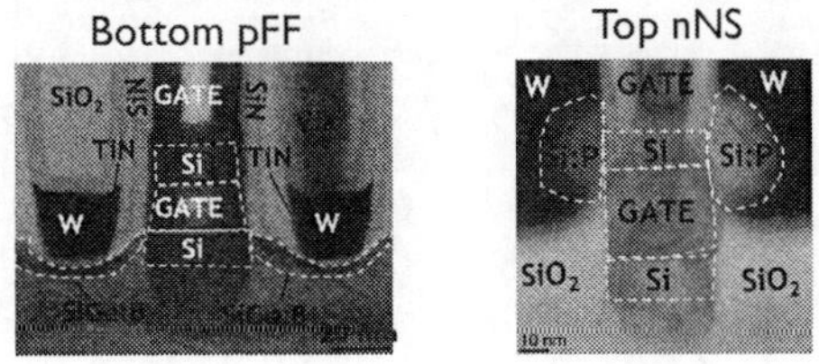

S. Subramanian et al VLSI2020

imec

15

CFET integration options

Monolithic CFET

STI w/ SiGe sac. layer

Bottom SD & contact

Top SD & contact

RMG

Sequential CFET

Bottom device process

Top ch. Layer transfer

Top device process

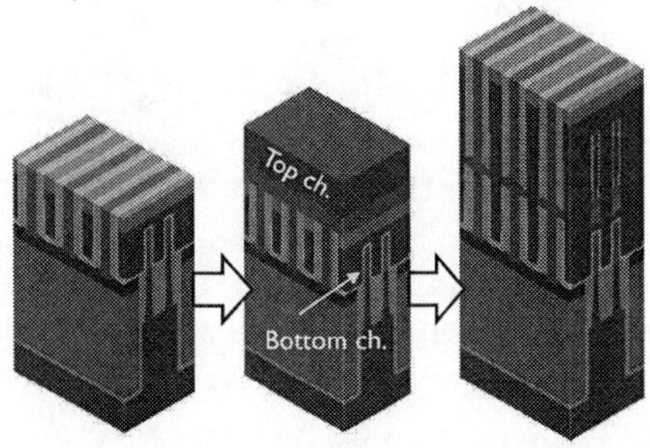

CFET integration options

- Monolithic: Bulk device integration from bottom to top
- Sequential: Bottom device process → Wafer bonding → Top device process

imec

16

Monolithic CFET Flow

S. Subramanian et al VLSI2020

Cost effective but need high ar patterning & complex vertical integ.

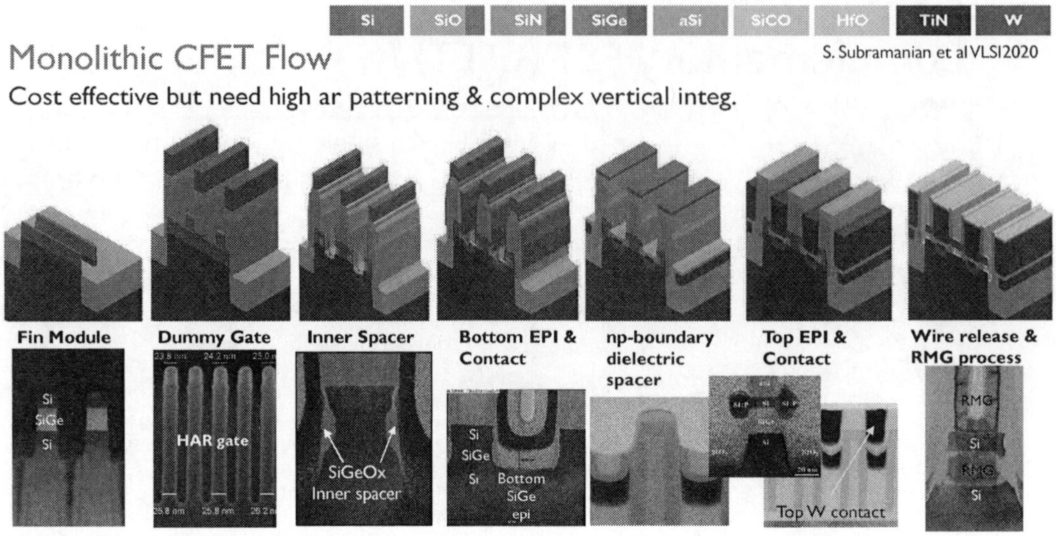

Challenges: Cleaning in high aspect ratio trench

Vertical edge placement control by precise CMP and etch back (see next slide)

imec

S. Subramanian et al VLSI2020

Monolithic CFET vertical edge placement control example

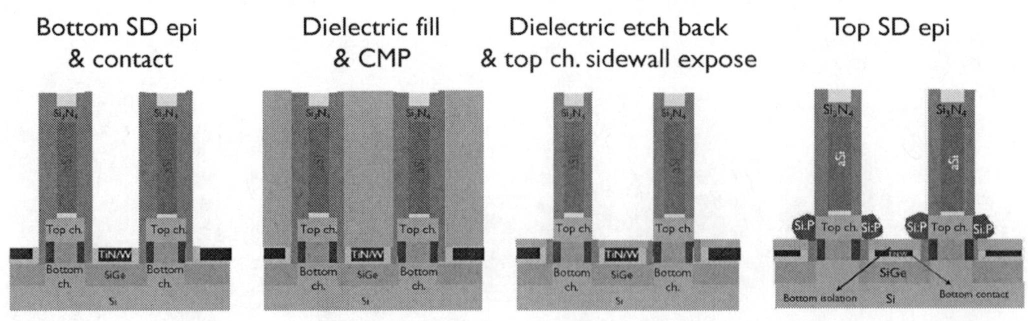

- In monolithic CFET integration, vertical edge placement controlled by CMP and etch back.
  - Precise CMP and etch back control is necessary.

imec

Sequential CFET flow

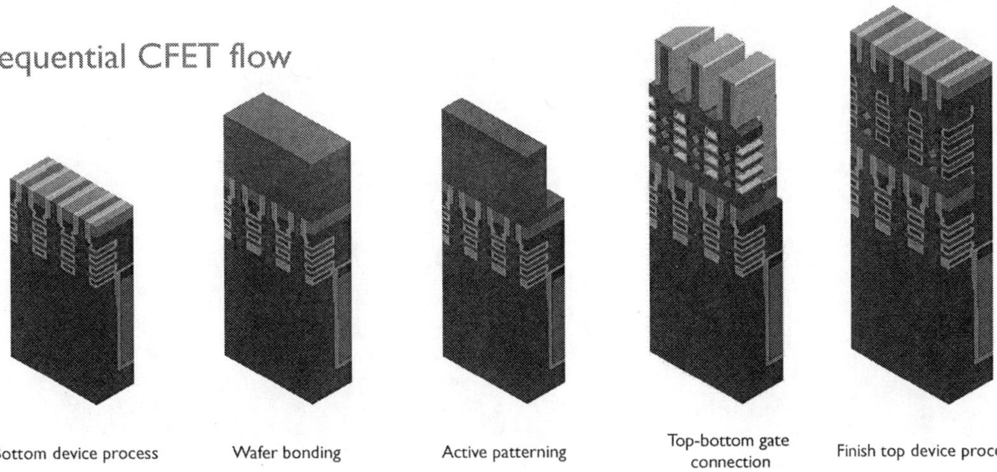

Challenge: Defect-free wafer bonding w/ scaled bonding oxide thickness (~30nm)
(Bonding oxide CMP uniformity & post CMP clean)

imec

19

Wafer bonding defect control

A.Vandooren et al IEDM 2018

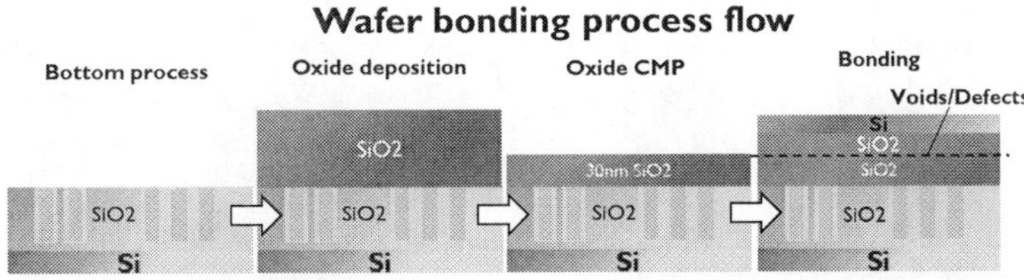

Patterned wafer geometry after bottom process

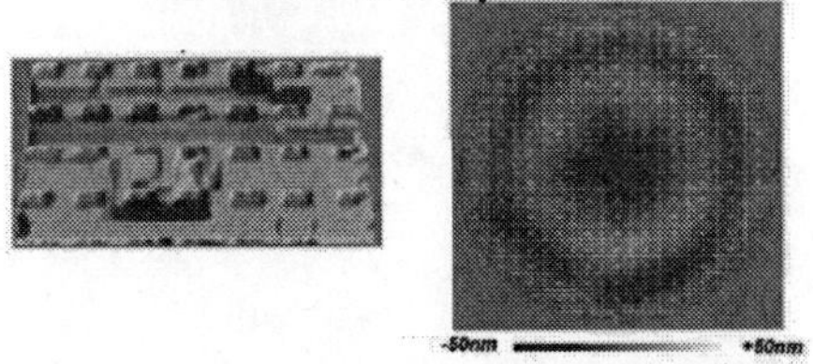

Bonding oxide thickness impact on bonding voids

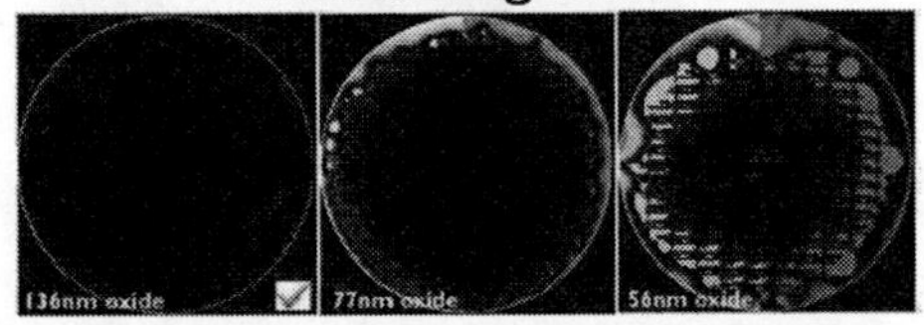

Underling topography induces bonding voids, especially at thin bonding oxide.

- Topography control by CMP is important

imec

Buried Power Rails (BPR)

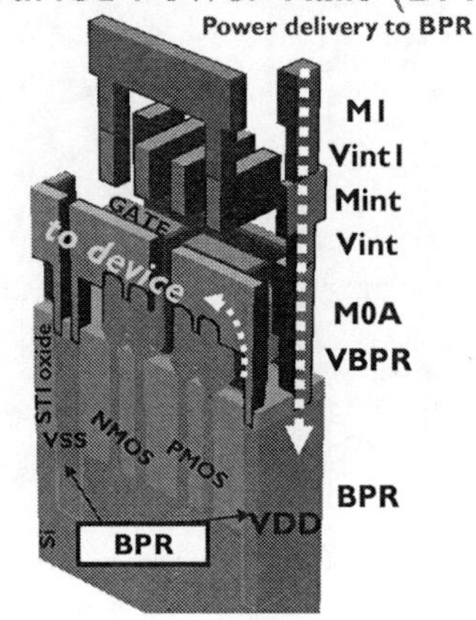

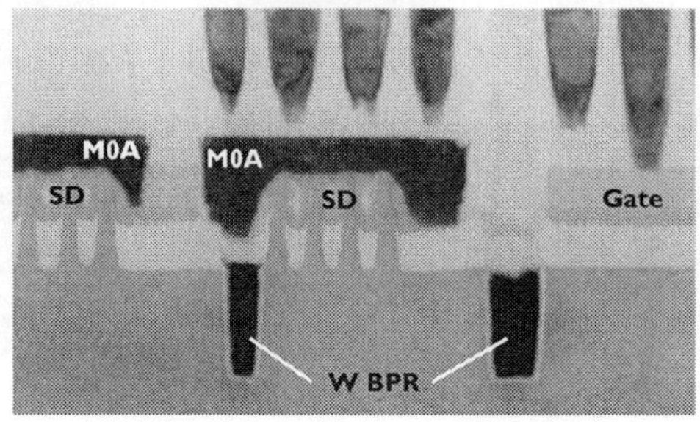

A. Gupta et al VLSI 2020

- BPR free up space in BEOL. → Area scaling.
- BPR line aspect ratio can be increased. → IR drop improvement

imec

21

BPR and VBPR process flow

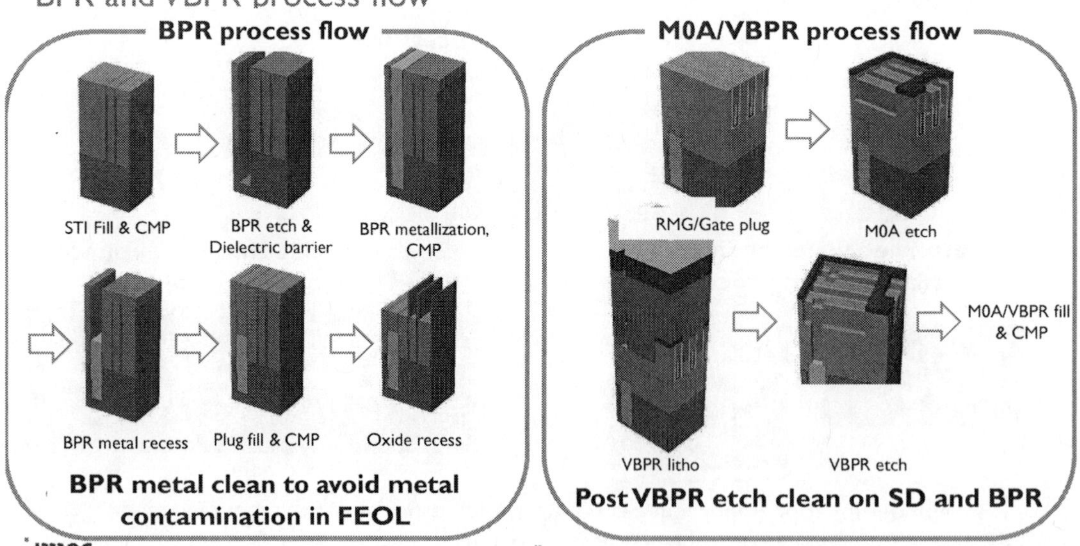

imec

22

BPR/VBPR metal candidates

BPR metal

	W	Ru
FEOL process compatibility		
Metal recess		
Electromigration		
Line Resistance		

VBPR metal

	Co	Mo	Ru
Compatibility w/ M0A			
Metal recess			
Electromigration			
Via Resistance			

A. Gupta et al VLSI 2020

- W and Co are first candidates for BPR and VBPR.
- Ru and Mo have potential to improve line and via resistance ➔ Need proper CMP/clean

Q.T. Le et al SPCC 2020

imec

23

Summary

- CMOS device scaling challenges
  - Track height (# of metal lines/standard cell) scaling to compensate pitch scaling slow down
    - Fin #/device reduction
    - N-P space reduction
    - Power line width scaling
- Device architecture evolution and its impact on surface preparation and clean
  - Nanosheet: Enables **single fin architecture** by wider device width/footprint than FinFET
    - SiGe & dielectric **high selective isotropic etch** is necessary for inner spacer & channel release. **Post-etch clean surface** is important for SD epi and gate stack.
  - Forksheet: Dielectric wall **reduces N-P space** than nanosheet.
    - **Dielectric wall isotropic selective etch** is additional step for FS. Dielectric wall loss has to be minimized in subsequent modules, such as SD epi and RMG.
  - CFET: **Smaller device CMOS footprint** than NS & FS, **N-P space is vertical** direction
    - Monolithic: **Post etch clean** in **high AR trench**. **Vertical edge placement control** by **precise CMP & etch back**.
    - Sequential: **Defect-free wafer bonding** with **thin bonding oxide** by **good CMP control** and **clean**
  - Buried Power Rail: **Power line area reduction** by high aspect ratio line buried in STI/substrate
    - **Proper clean** is necessary to integrate **BPR metal w/o contamination issues**.
    - **CMP/clean** for **new low resistive/barrier-less metal**, such as Mo and Ru, needs to be developed for the future.

imec

24

embracing a better life

PUBLIC

UNISERS
UNIVERSAL CONTAMINATION ANALYZER

Sub-10nm defect inspection of unpatterned wafers with UNISERS and application to liquid particle analysis

altun@unisers.ch
+41 789339260
www.unisers.ch

Ali Ozhan Altun, Ph.D.

SPCC 2020

Outline

- Introduction
- Performance characterization for single particle detection
  - Characterization of the detection limit
  - Material-independent size estimation & dynamic range
  - Scanning options
- Liquid defectivity analysis with successive spin-drying
- Conclusions

Introduction:

- Enhancement of the polarization field of particles deposited on wafer:
  - Enhanced Raman scattering-> Speciation (SPCC 2019)
  - Enhanced total scattering-> Detection (particle size estimation)

RECAP (SPCC 2019, Portland)

- Speciation of non-volatile impurities (carbon & silica)

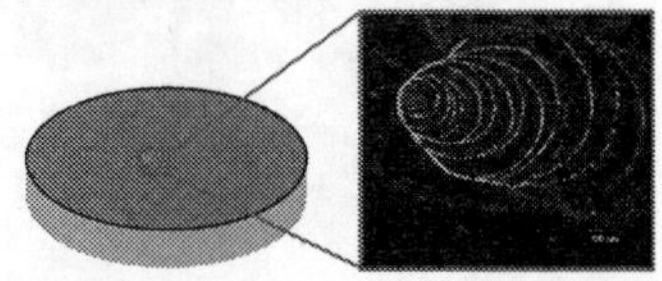

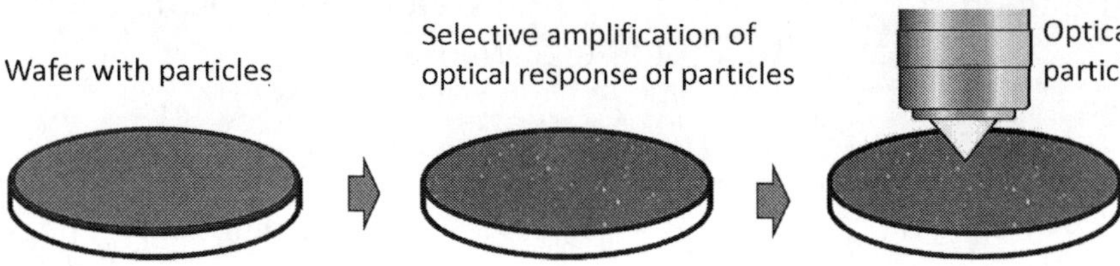

www.unisers.ch

Performance characterization methodology

1. Identification of the mean particle size of commercially available particles
   - Center-to-center distance measurement
2. Characterization of the UNISERS detection efficiency based on standard silicon particles
   - SEM-UNISERS correlation method
3. Characterization of the size estimation performance
   - Size distribution histograms

www.unisers.ch

How to characterize the mean particle diameter with SEM images?

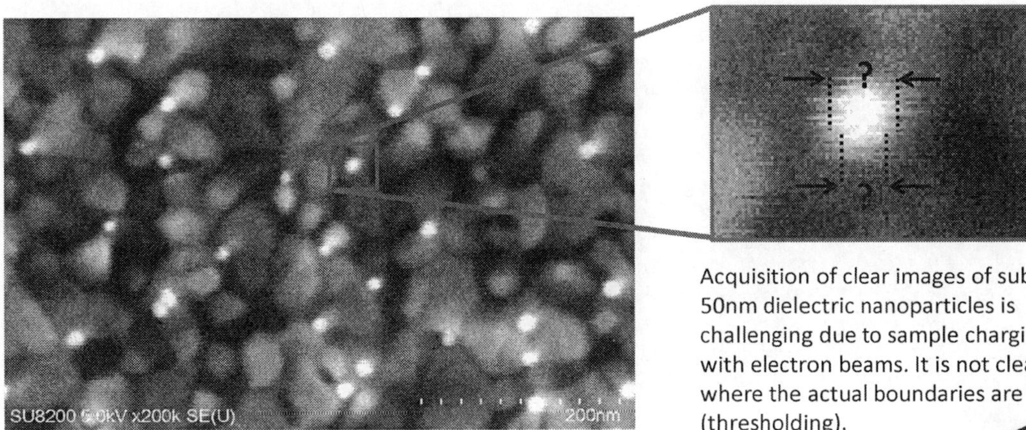

Acquisition of clear images of sub-50nm dielectric nanoparticles is challenging due to sample charging with electron beams. It is not clear where the actual boundaries are (thresholding).

www.unisers.ch

5

Detection of center-to-center distances

Center-to-center distances are robust to charging effects. Calculation of center-to-center distances of trimers in SEM images is a reliable method to determine **the upper bound of the "mean particles diameter"**.

A trimer of nanoparticles.

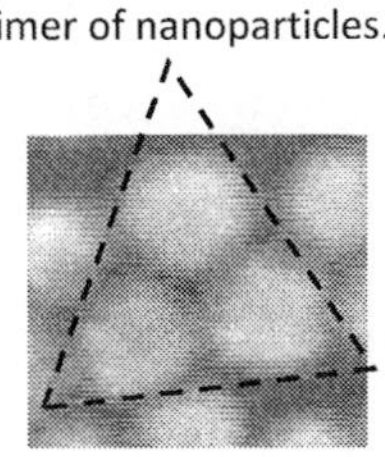

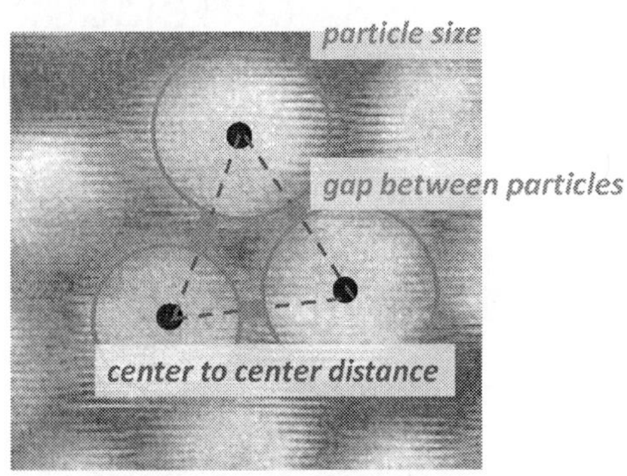

$$\textit{Average particle size of trimer} + \textit{gap between particles} = \frac{\sum \textbf{center to center distance}}{\mathbf{3}}$$

www.unisers.ch

6

Characterization of mean diameters based on center-to-center distances

Some trimers used to detect center-to-center distances

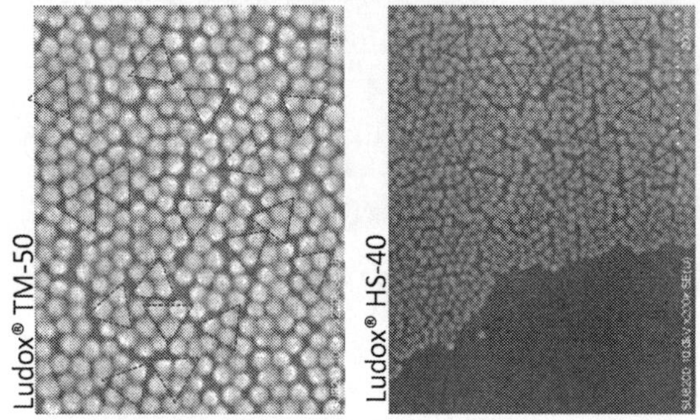

- Colloidal crystals can be created by spin-coating high-concentration colloidal liquid sample.

Particle name	Particle type	Detected mean diameter	Producer's specification
Ludox TM50	Silica	**29.7 nm**	22 nm
Ludox HS40	Silica	**16.3 nm**	12 nm
Ludox SM	Silica	**8.9 nm**	7 nm
Si150	Silica	**134 nm**	150 nm
PS31	Polystyrene	**23.5 nm**	31±3 nm
PS23	Polystyrene	**16.2 nm**	23±3 nm
Au31	Gold	**32.1 nm**	31 nm
Au60	Gold	**56.3 nm**	60 nm

Data used for UNISERS calibration

www.unisers.ch

7

Characterization of the detection limit with SEM-UNISERS correlation method

Spike Si-wafer with standard silica particles (e.g. Ludox) by spin-drying

SEM image acquisition on certain coordinates

UNISERS image acquisition on identical coordinates

Particle-to-particle correlation of SEM and UNISERS images using particle deposition patterns

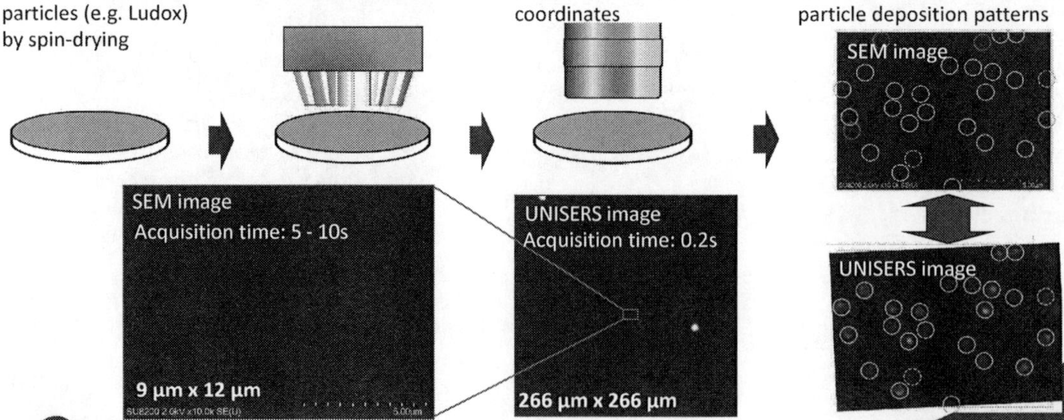

www.unisers.ch

8

Example SEM-UNISERS correlation for Ludox TM50 on silicon wafer (mean particle size: 29.7nm)

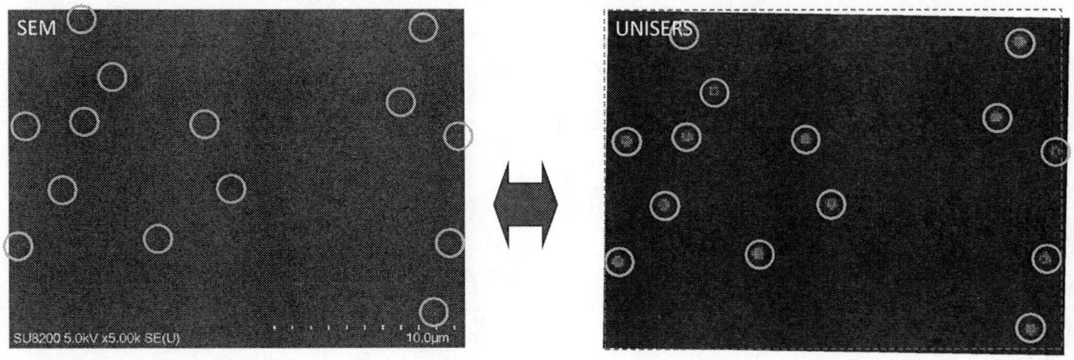

14 particles detected
0 particle not detected
0 particle not visible with SEM

www.unisers.ch

9

Example SEM-UNISERS correlation for Ludox TM50 on Au-coated wafer (mean particle size: 29.7nm)

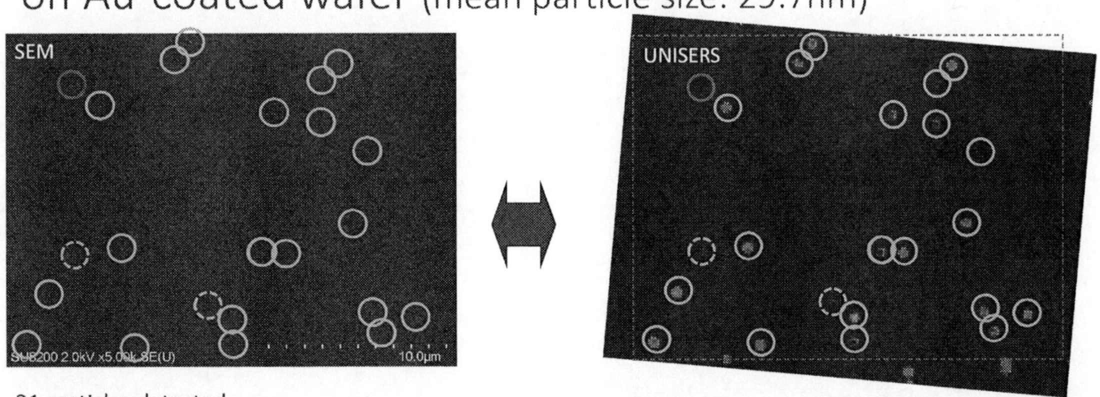

21 particles detected
1 particles not detected
2 particles not clearly visible with SEM*

*Surface roughness of Au film makes it difficult to discriminate a very small silica particle on SEM images

www.unisers.ch

10

Example SEM-UNISERS correlation for Ludox HS40 on silicon wafer (mean particle size: 16.3 nm)

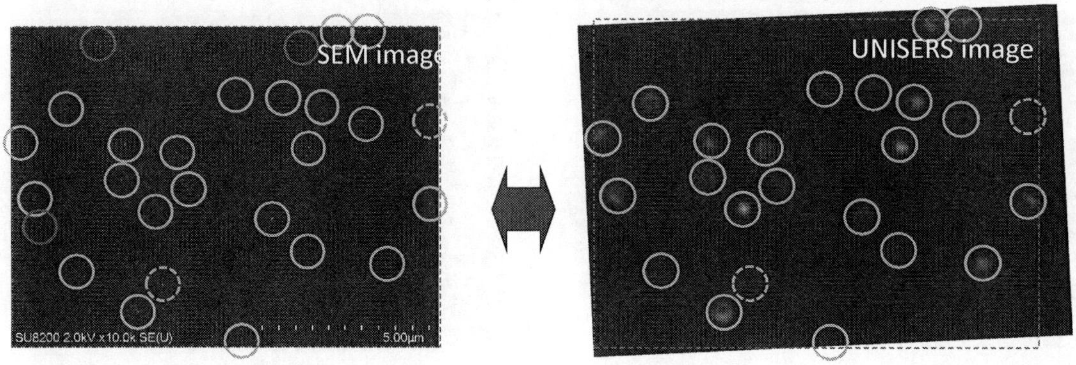

22 particles detected with 100% probability
2 particles are detected with 50% probability*
3 particles not detected
0 neighbor particles are ignored

*Depends on the haze (airborne molecular contamination level)

www.unisers.ch

11

SEM-UNISERS correlation test result

Particle type	Bottom substrate	Number of SEM/UNISERS images	Total single particles imaged with SEM	Cleary detected single particles with UNISERS	Detected with 50% probability	Not detected
Ludox TM50 Silica, 29.7 nm	Bare silicon	4	60	60	0	0
	Au-coated silicon	4	108	99	0	9
Ludox HS40 Silica, 16.3nm	Bare silicon	16	312	242	42*	28

*Very small particles can be detected if haze (background) signal is very low.

Detection efficiency of TM50: 90%-99%
Detection efficiency of HS40: 75%-90%

www.unisers.ch

12

Example zoom-in UNISERS images

Ludox SM (8.9nm)

Ludox HS-40 (16.3nm)

Ludox TM-50 (29.7nm)

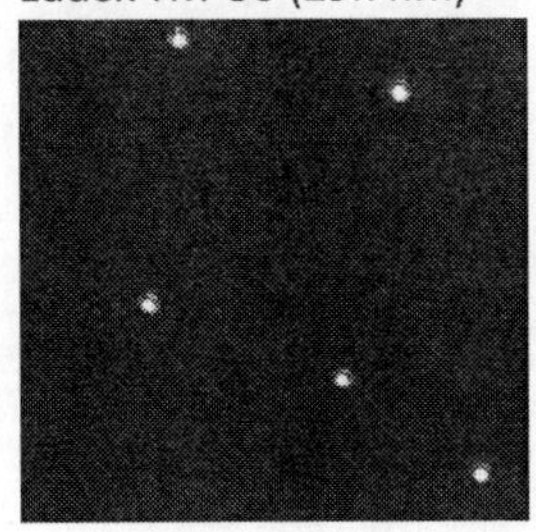

Threshold for particle detection: (a) Intensity (b) Pixel count

Example1: Particle size distribution of Ludox TM50

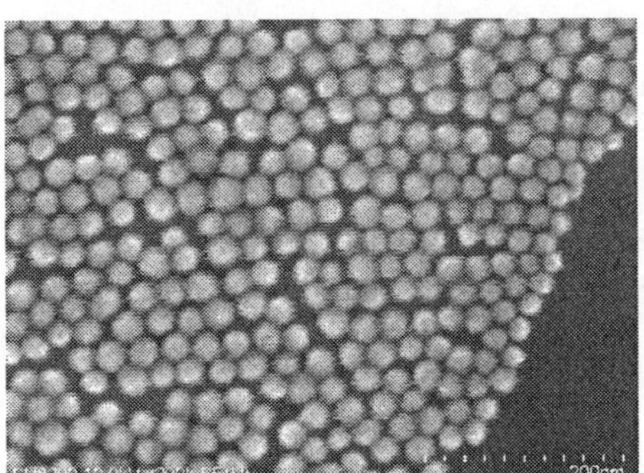

UNISERS ANALYSIS RESULT

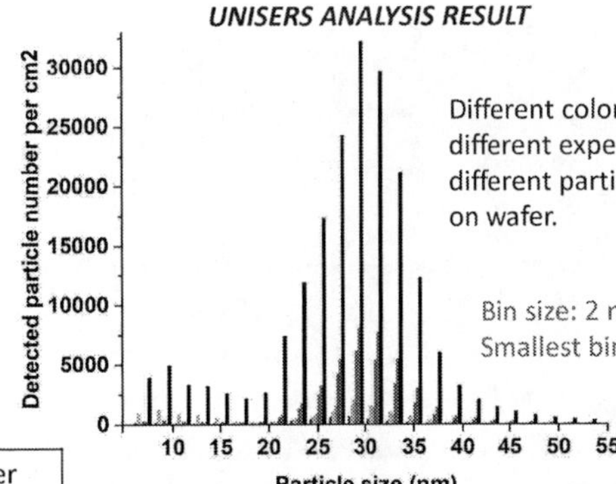

Different colors represent different experiment with different particle density on wafer.

Bin size: 2 nm
Smallest bin: 7nm-9nm

Ludox TM50	Mean particle diameter
Producer's specification	22 nm
SEM Center-to-center analysis	29.7 nm
UNISERS Analysis*	29.3 nm

- UNISERS analysis is performed on individual particles deposited on the wafer (not closely packed colloidal crystals)

*Mean particle diameter is calculated with Gaussian curve fitting

Example 2: Particle size distribution of Ludox HS40

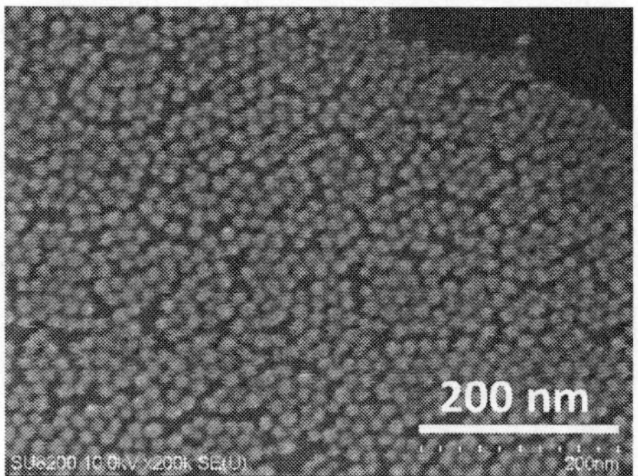

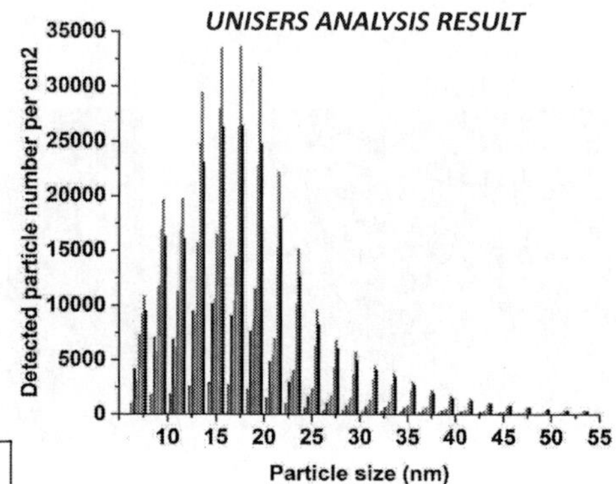

Ludox HS40	Mean particle diameter
Producer's specification	12 nm
SEM Center-to-center analysis	16.3 nm
UNISERS Analysis	16.0 nm

www.unisers.ch

15

Example 3: Particle size distribution of Ludox SM

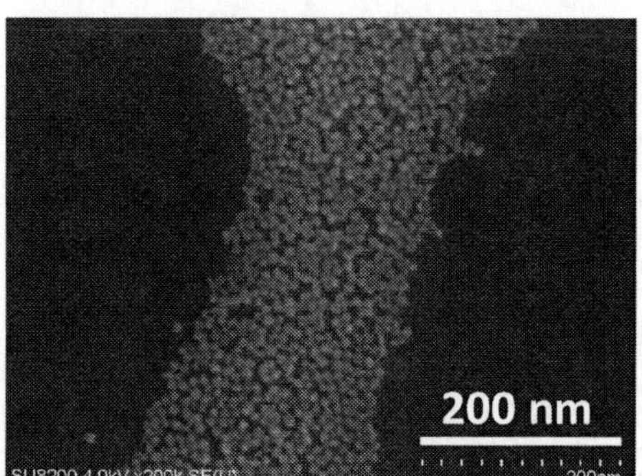

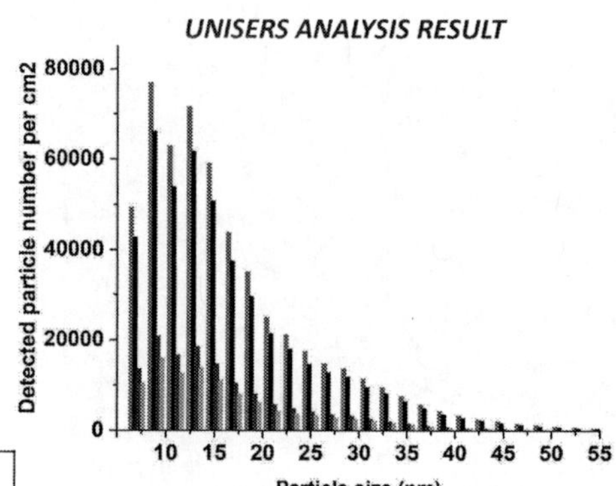

Ludox SM	Mean particle diameter
Producer's specification	7 nm
SEM Center-to-center analysis	9.3 nm
UNISERS Analysis	11.1 nm

www.unisers.ch

16

Example 4: Particle size distribution of PS31

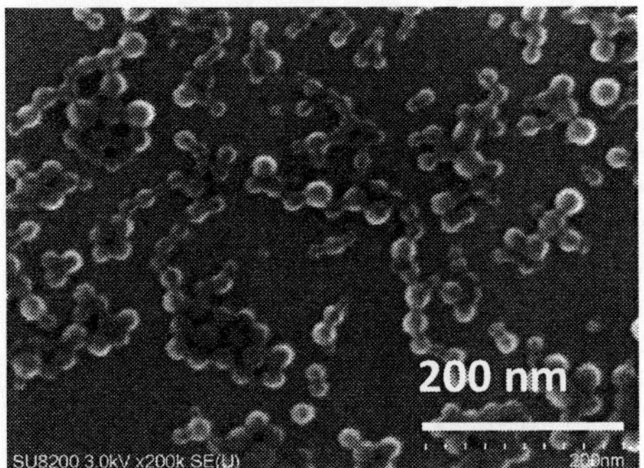

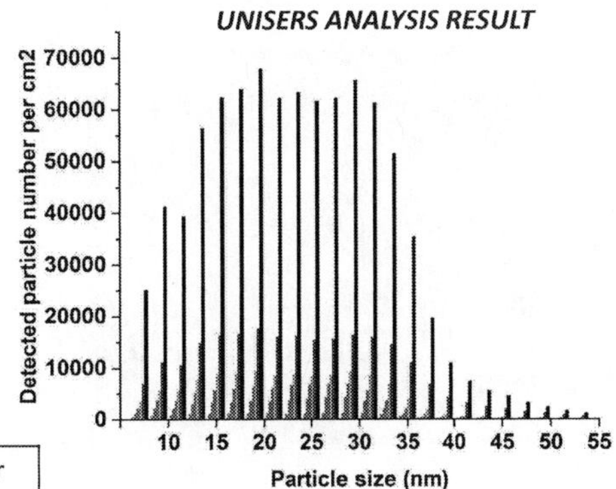

PS31	Mean particle diameter
Producer's specification	31±3 nm
SEM Center-to-center analysis	23.5 nm
UNISERS Analysis	22.1 nm

- Much wider particle size distribution compared to Si or Au

Example 5: Particle size distribution of PS23

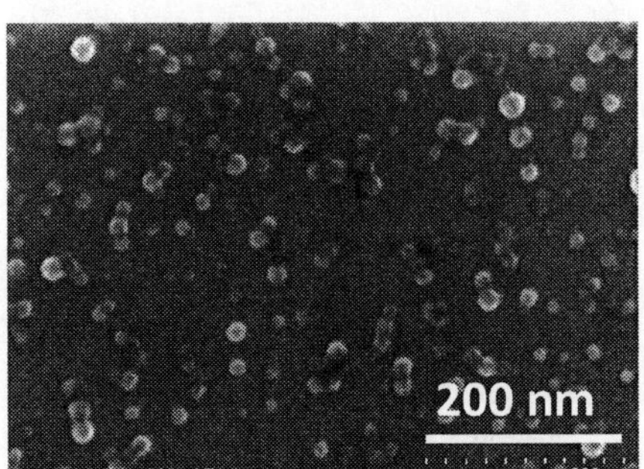

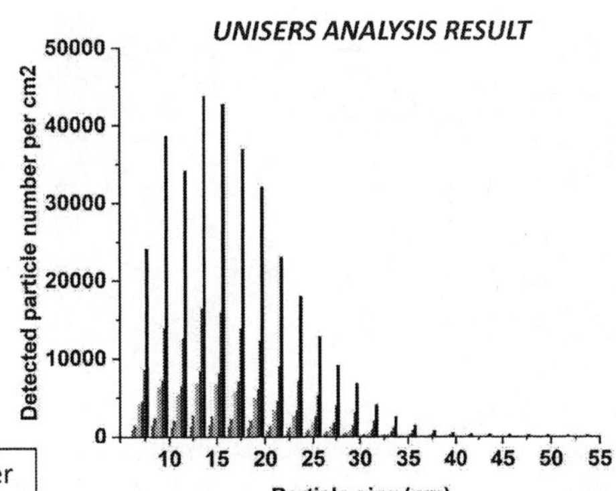

PS23	Mean particle diameter
Producer's specification	23±3 nm
SEM Center-to-center analysis	16.2 nm
UNISERS Analysis	14.9 nm

Example 6: Particle size distribution of Au31

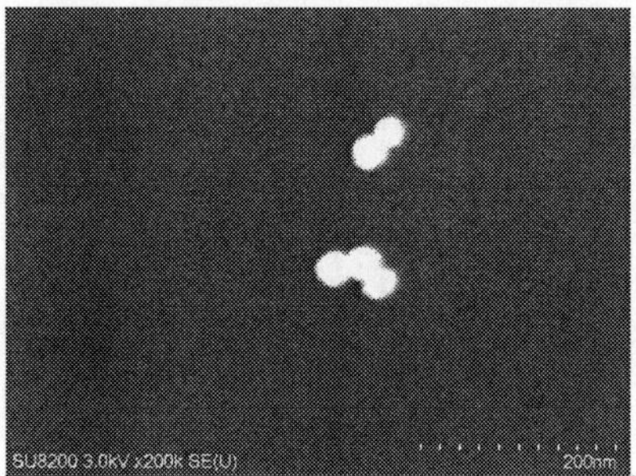

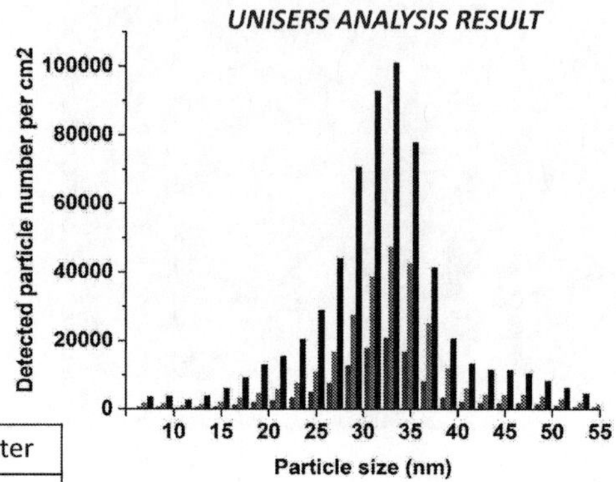

Au31	Mean particle diameter
Producer's specification	31 nm
SEM Center-to-center analysis	32.1 nm
UNISERS Analysis	32.2 nm

www.unisers.ch

19

Example 7: Particle size distribution of Au60

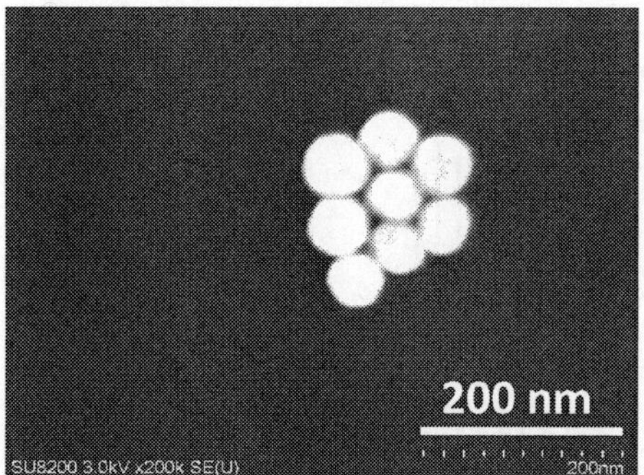

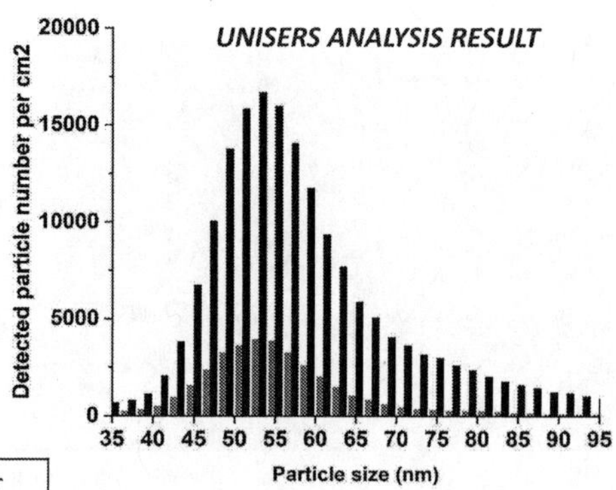

Au60	Mean particle diameter
Producer's specification	60 nm
SEM Center-to-center analysis	56.3 nm
UNISERS Analysis	54.2 nm

www.unisers.ch

20

Example 8: Particle size distribution of Si150

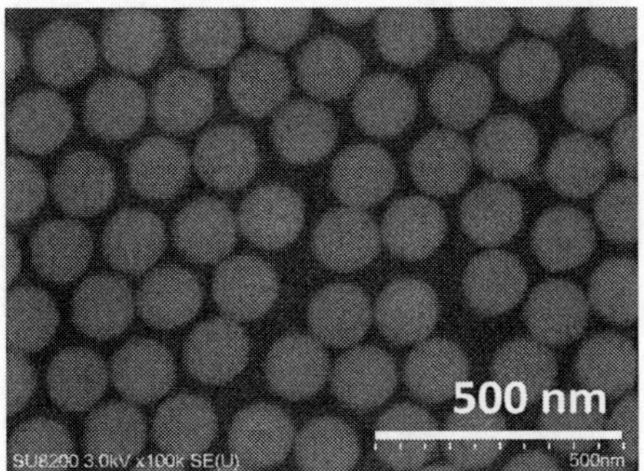

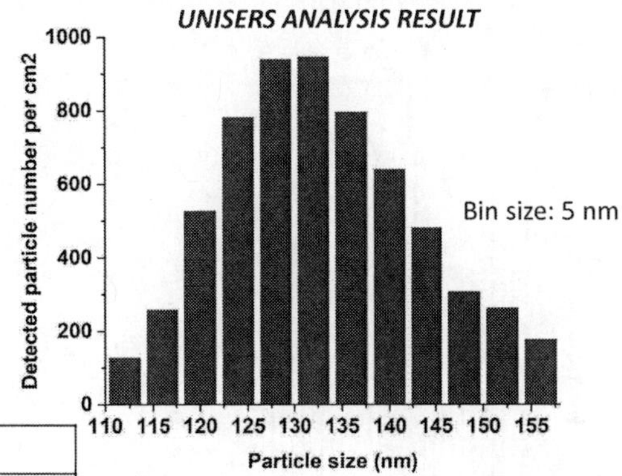

Silica150	Particle size (nm)
Producer's specification	150 nm
SEM Center-to-center analysis	134 nm
UNISERS Analysis	133 nm

www.unisers.ch

21

Example 9: Particle size distribution of Au5

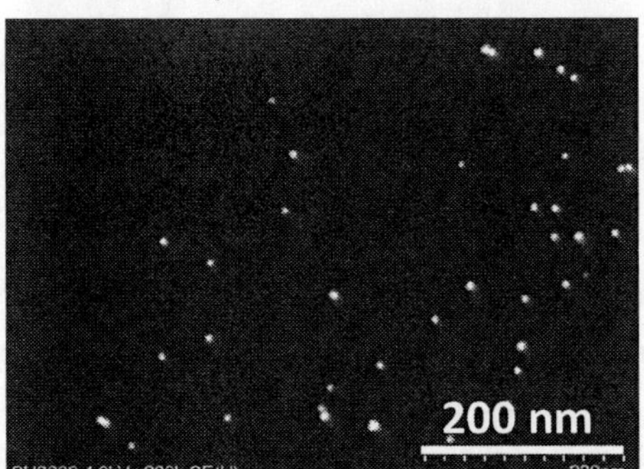

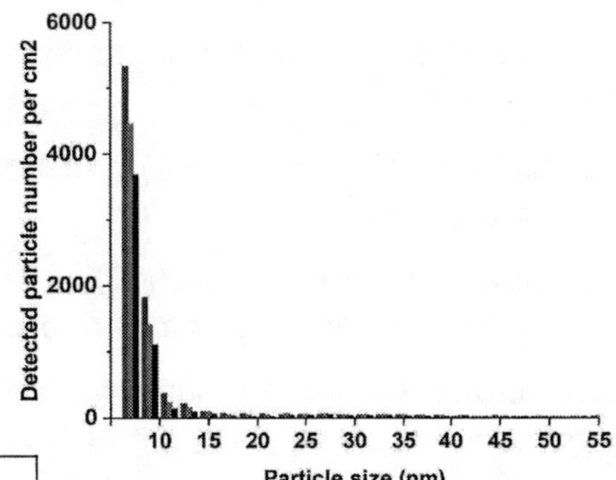

Au5	Particle size (nm)
Producer's specification	5 nm
SEM Center-to-center analysis	Could not performed
UNISERS Analysis	7 nm

www.unisers.ch

22

Material-independent size estimation

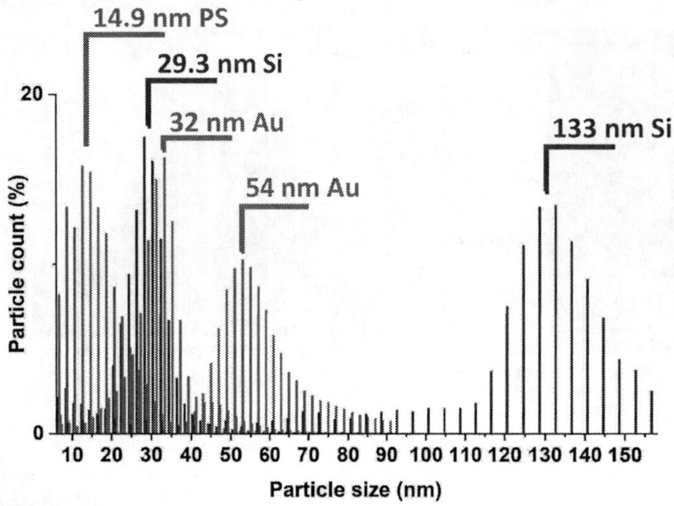

- Material-independent size characterization over a large dynamic range (with a single optical scan)
- Check appendices for more particle size characterization data

www.unisers.ch

General overview of UNISERS Particle detection technology

- Particle detection limit: sub-10nm with a low light budget. Signal amplification for 20-nm particle is ~10^6.
- Detection limit is valid even on high-scattering backgrounds (e.g. metal-coated substrates).
- Particle size estimation with (1) signal intensity & (2) pixel count: Large dynamic range (sub-10nm to 150 nm with a single scan).
- 2-micron resolution for 30nm particles (can distinguish two different particles if they are located with a minimum distance of 1 micron)
- Material-independent size estimation (true size estimation)
- 20mins/cm^2 scan speed for sub-10nm sensitivity.

www.unisers.ch

Application 1: Liquid particle analysis with successive spin-drying

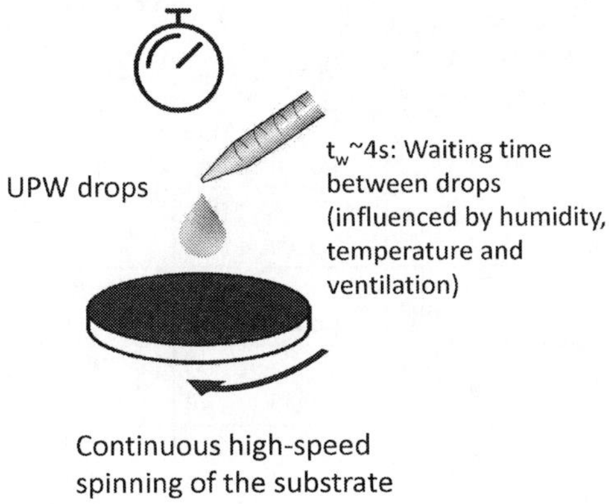

Concept of successive spin-drying and particle deposition:

Particle deposition to the substrate happens:

1. during initial contact to the substrate and
2. after evaporation of UPW from the substrate

Repeating this process on the UNISERS substrate increases the particle density. Experiments in this document show the increase of the particle density is linear with the applied drop number. For example, applying 100 successive drops increases the particle density on the substrate by x100. Increased particle density on the surface means improved detection limit for particle concentrations in liquids.

www.unisers.ch

Particle deposition results for Ludox HS40®

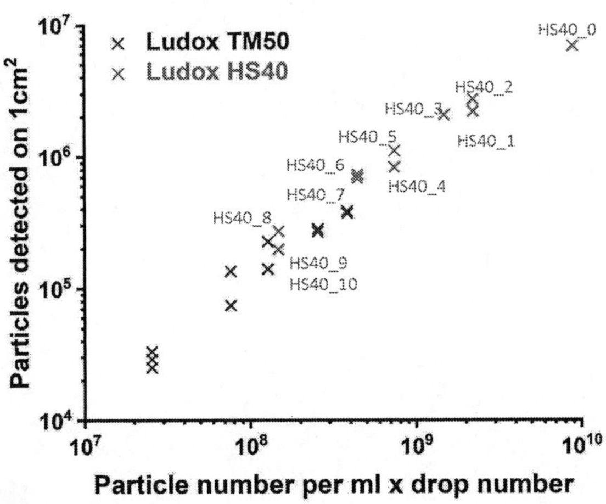

Ludox®-HS40 silica particle (16nm) spiked samples		
Sample ID	***Concentration (ppt, Si)***	***Drop number***
HS40_0	240	100
HS40_1	60	100
HS40_2	40	150
HS40_3	40	100
HS40_4	40	50
HS40_5	20	100
HS40_6	40	30
HS40_7	12	100
HS40_8	40	10
HS40_9	4	100
HS40_10	4	100

www.unisers.ch

Quantification of liquid particle concentration

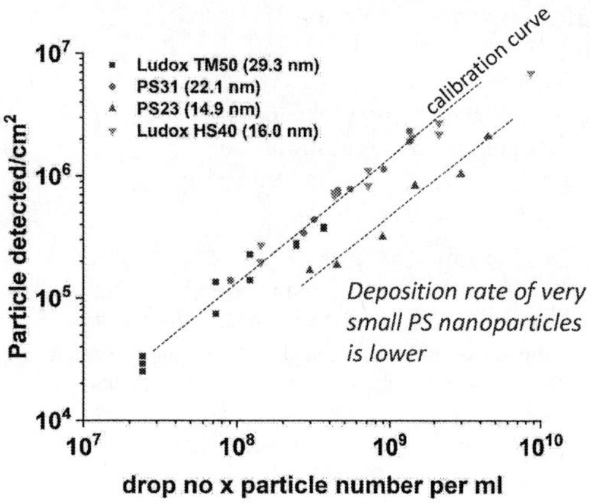

Drop number	*Total time required for successive spin-drying*	*Particle concentration of UPW*	*Detected particles (per cm2)*
1	1 min	1'000 particles/ml	1
10	2 min	1'000 particles/ml	10
100	8 min	1'000 particles/ml	100
1'000*	70 min	1'000 particles/ml	1'000
10'000*	12 hours	1'000 particles/ml	10'000

*Extrapolated data

www.unisers.ch

27

Conclusions

❖UNISERS is a sub-10nm optical defect scanner.

❖Currently, we can analyze up to 4-inch wafers or wafer pieces.

❖Defect speciation with enhanced Raman in progress.

www.unisers.ch

28

Don’t hesitate to contact us with open questions, proposals for joint projects or collaborations

CEO Dr. Ali O. Altun
altun@unisers.ch
+41 78 933 92 60

CTO Fabian Walter
walterf@unisers.ch
+41 79 420 05 02

UNISERS Ltd.
c/o ETH Zurich ieLab, HPL D16.2
Otto-Stern-Weg 7
8093 Zurich
Switzerland

Spinoff
ETH zürich

WET-CHEMICAL CLEANING OF COBALT AND MOLYBDENUM FOR ADVANCED INTERCONNECTS

Q.T. LE, E. KESTERS, A. USMAN IBRAHIM*, A. KLIPP*, H. GUEVENC*,
A. PACCO, M. VAN DER VEEN, E. ALTAMIRANO SANCHEZ, AND F. HOLSTEYNS
IMEC, LEUVEN, BELGIUM
* BASF SE, LUDWIGSHAFEN, GERMANY
Email address: QuocToan.Le@imec.be

SPCC, September 1-2, 2020

PUBLIC

Outline

- Introduction
  - Cobalt as an alternative metal to Copper
  - Co etching: commodity vs. formulated chemicals
- Co cleaning using Fotopur RD-F mixture
  - Experimental: Cleaning chemistry and objectives
  - SEM, XPS, and TEM characterization of patterned stack structure
  - Electrical evaluation
  - Summary
- Mo cleaning: choice of chemicals
  - Effect of annealing
  - Effect of H_2O_2 as oxidizer concentration (H_2O_2)
  - Summary

2

INTRODUCTION

COBALT: INTERCONNECT ALTERNATIVE TO COPPER

Line resistance trends for Cu vs. Co interconnects under certain assumptions

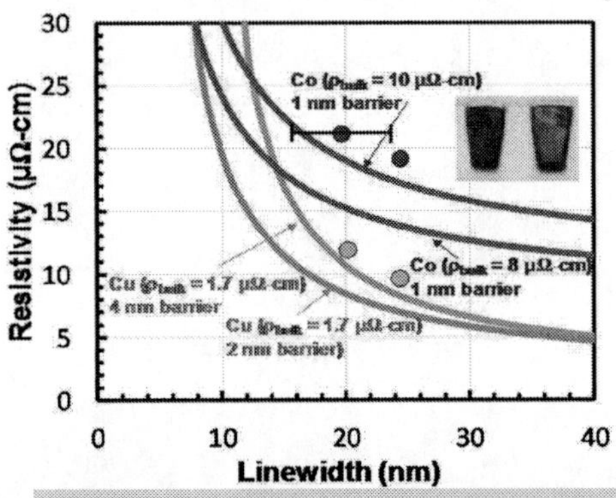

1. J. Kelly *et al.*, and V. Kamineni *et al.*, IEEE IITC/AMC 2016.
2. D. C. Edelstein, IEDM 2017.

- Alternative metal(s) to mitigate Cu resistance scaling impacts for lines and vias, and from size-effects
- Co: interconnect metal
  - higher bulk resistivity, but lower size-effects and requires ultrathin barrier
  - Co vs. Cu: resistance crossover may occur below ~15 nm CD

imec BASF

Co ETCHING: COMMODITY VS. FORMULATED CHEMICALS

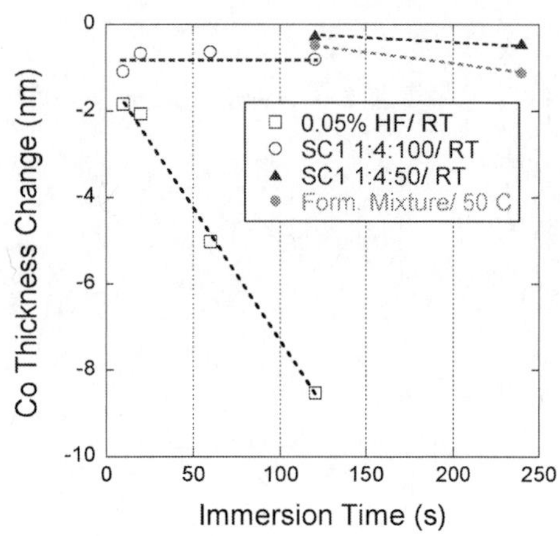

- Co etch rate
  - 0.05% HF > Form. chem. with HF pre-treat > SC1 1:4:100 ~ SC1 1:4:50 ~ Form. chem.
  - In acidic medium

$$Co + H_2O \rightarrow Co(H_2O)_{ads} \rightarrow Co\,(OH)^+ + H^+$$

$$\downarrow \; Co^{2+} + H_2O$$

- Characteristics

	Commodity	Formulation
Choice	Large	Limited choice
Co compatibility	Low	High (due to appropriate pH, additives)
Etch residue removal	Limited efficiency	Higher

More info on Co etch/HF:

1. L. Broussous *et al.*, UCPSS 2016.
2. Y. Akanishi et al., UCPSS 2018.

imec BASF

Co CLEANING USING FOTOPUR RD-F MIXTURE

BASF
We create chemistry

CLEANING CHEMISTRY AND OBJECTIVES

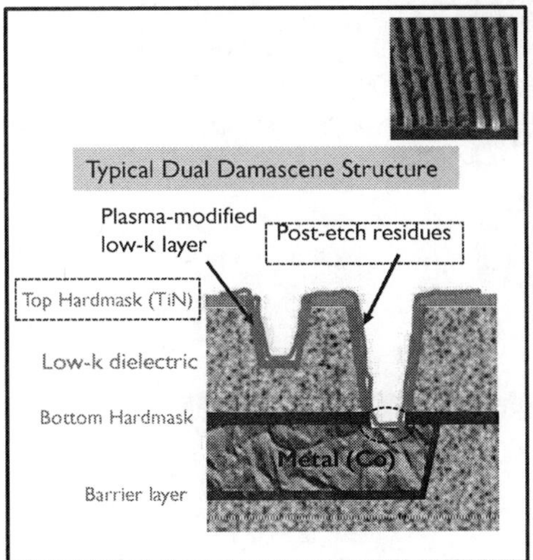

Three main objectives:

- TiN hardmask: to be removed
- Post-etch residues: to be removed before metallization
- No metal etch & no oxidation

CLEANING CHEMISTRY

- Fotopur RD-F (BASF)
  - RD-F:H_2O_2 = 1:1/ 50 °C

TEST VEHICLES

- Blanket films (Co, TiN, OSG)
- Patterned structures

imec BASF

MATERIAL COMPATIBILITY AND TiN ETCH BY RD-F

Plasma-treated OSG 2.85

	Pre clean		Post clean	
	Thickness (nm)	RI @ 633 nm	Thickness (nm)	RI @ 633 nm
Mean	53.2	1.3880	53.0	1.3891
Min	50.5	1.3871	50.0	1.3883
Max	55.2	1.3886	55.0	1.3898

ECD Co

	Rs_Pre (Ohm/sq.)	Rs_Post (Ohm/sq.)
Mean	0.284	0.282
Stdv	0.021	0.012

TiN etch

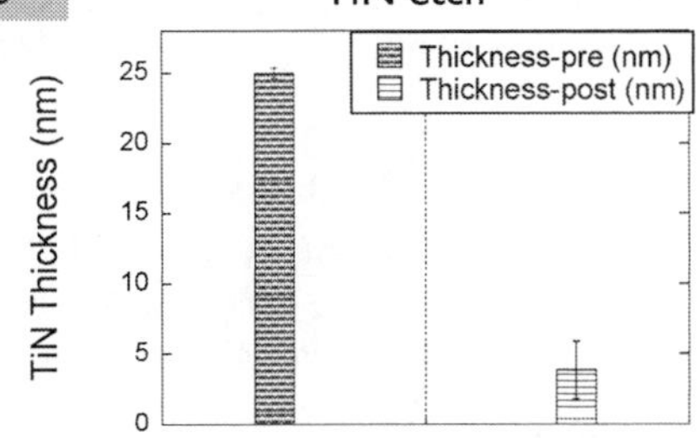

- RD-F:H_2O_2 1:1 @ 50 °C/ 2 min
- Good compatibility with OSG 2.85 and ECD Co
- TiN etch: ~3 nm remained for 2 min clean
  - Artifact (SE measurement)
  - Complete TiN etch confirmed using patterned structure

imec BASF

PERR EFFICIENCY: X-SEM AND XPS RESULTS

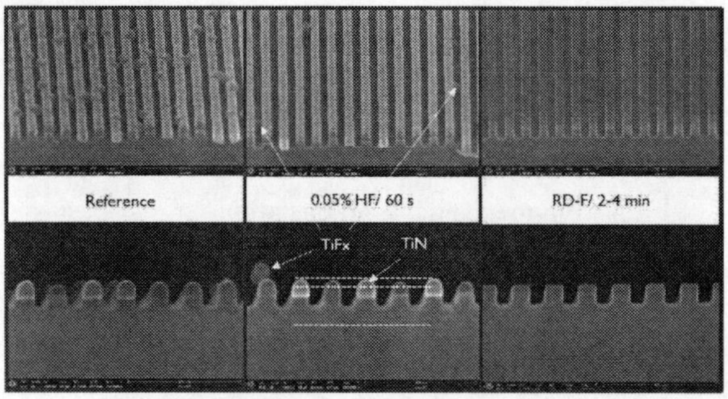

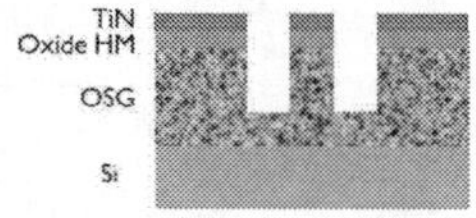

- RD-F:H_2O_2 1:1 @ 50 °C
  - TiN HM removed together with TiFx and CFx residues
  - No LK dielectrics etch after 4 min

Atomic concentrations (At.%)

	C	O	Si	Ti	N	F
REF	26.4	27.4	8.0	8.3	6.1	23.8
0.05% HF	28.2	24.9	10.7	8.3	10.6	17.3
2min RD-F	15.9	53.5	26.1	0.1	1.1	3.3
4min RD-F	16.2	53.6	26.2	0.1	1.2	2.7

imec BASF

XPS CHARACTERIZATION OF PATTERNED OSG STACK

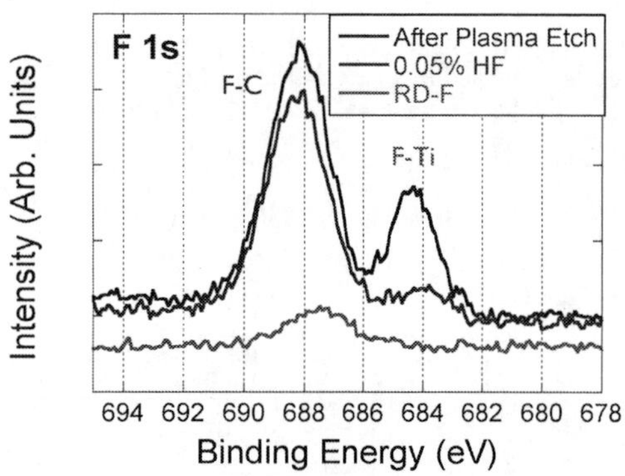

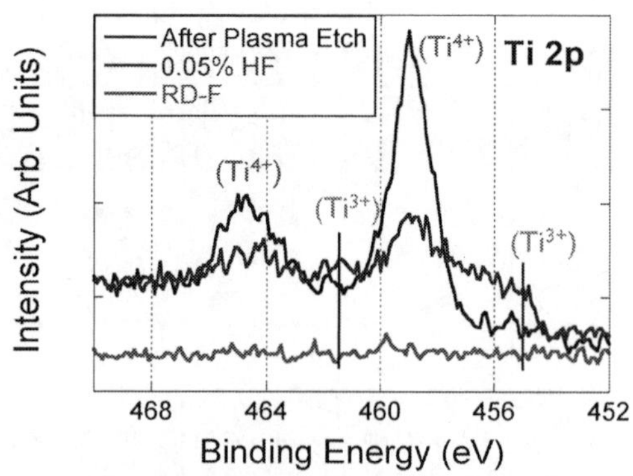

- Both TiFx residues and TiN HM were completely removed in RD-F:H_2O_2 mixture
- Small amount of fluorinated residues (detected at BE ~688 eV) still remained, which could be understood by incorporation of some residues into the OSG pores during dry dielectric etch

imec BASF

Co ETCH AND GALVANIC CORROSION

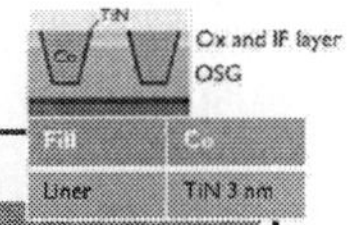

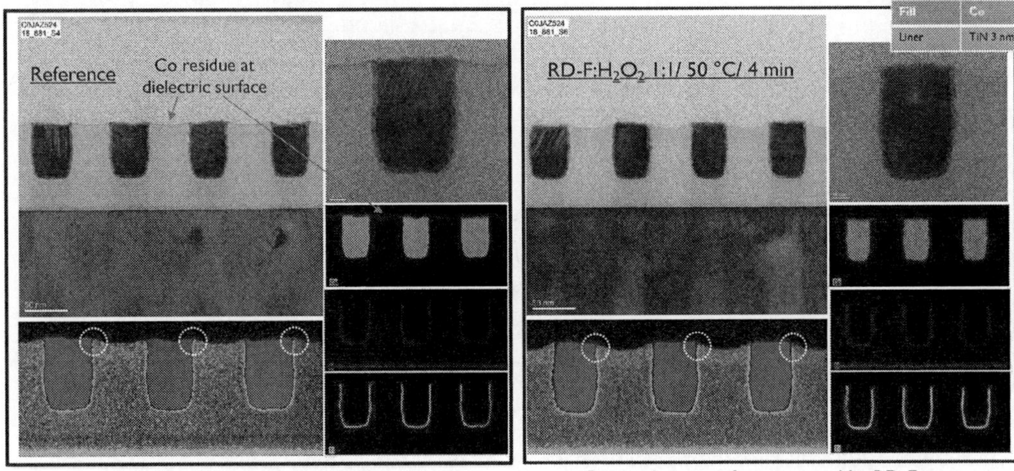

- Presence of Co residue on SiO_2 surface
- Some dishing already observed after CMP

- Co residue at surface removed by RD-F
- No Galvanic corrosion induced by RD-F chemistry

imec

ELECTRICAL EVALUATION: EXPERIMENTAL

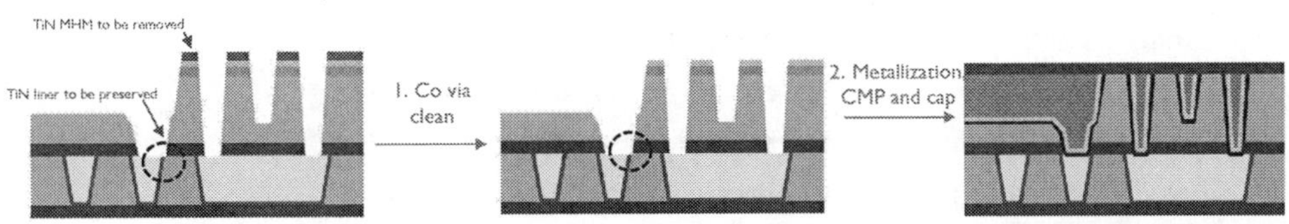

22 nm HP Cu Metal-2 on Co metal-1

Objective: Co clean evaluation using RD-F mixture, on 22 nm HP dual damascene structure

EXPERIMENT DETAIL:

#	Co Clean	Rinse
Co1	2 min RD-F	Aqueous alkaline + IPA rinse
Co2	2 min RD-F	Aqueous alkaline rinse
Co3	5 s dHF + 2 min RD-F	Aqueous alkaline + IPA rinse
Co4	5 s dHF + 4 min RD-F	Aqueous alkaline + IPA rinse

imec BASF

KELVIN VIA RESISTANCE

#	Co Clean	Rinse
Co1	2 min RD-F	Aqueous alkaline + IPA rinse
Co2	2 min RD-F	Aqueous alkaline rinse
Co3	5 s dHF + 2 min RD-F	Aqueous alkaline + IPA rinse
Co4	5 s dHF + 4 min RD-F	Aqueous alkaline + IPA rinse

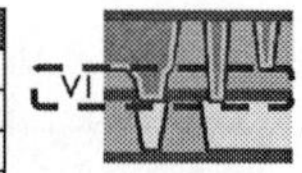

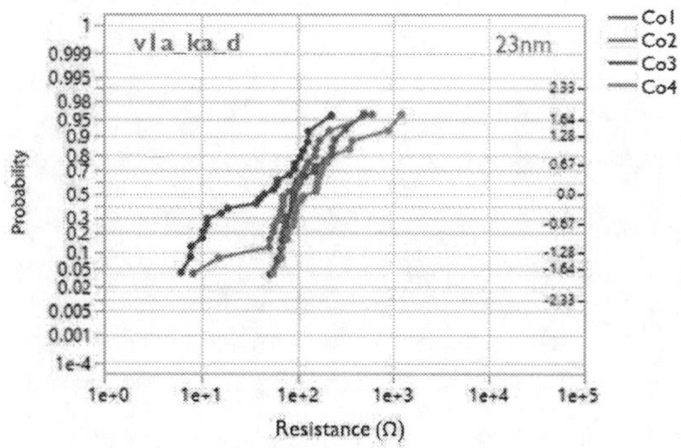

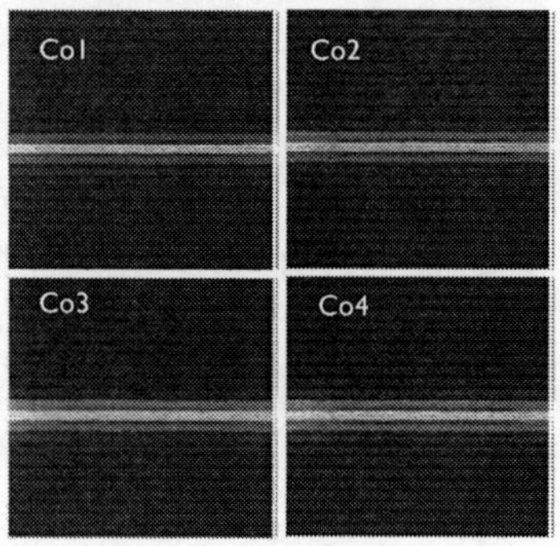

TD-SEM after CMP

- dHF pre-treatment (Co3 and Co4) ➔ lower resistance but larger distribution
- Only a small difference between Co1 and Co2 where Co1 seems to be slightly better
  - The aqueous alkaline rinse followed by IPA-rinse (Co1) is better than without (Co2)

imec BASF

13

SUMMARY

- Co cleaning using Fotopur RD-F cleaning chemistry
  - Good residues removal (both TiFx and organic residues) and TiN HM etch
  - No Galvanic corrosion induced by RD-F chemistry
  - 22 nm MP structure: good electrical performance (via resistance)
    - dHF pre-treatment of 5 s: appears to have a noticeable impact on resistance (slight etching of damaged LK layer)
      ➔ larger resistance distribution
    - An IPA rinse is necessary to obtain lower resistance

imec BASF

14

Mo CLEANING: CHOICE OF CHEMICALS

OBJECTIVE AND EXPERIMENTAL

Objective: Mo compatibility with chemical mixtures

- Mo samples: deposited by PVD
  - Nominal thickness: 50 nm
  - As-deposited and annealed at 420 and 650 °C
  - Annealing atmospheres: FG, N_2, and H_2
- Choice of chemical mixtures
  - *0.05% HF*
  - *SC1 mixture with variable [H_2O_2]*
  - *Formulated mixture, pH >12*
  - *Formulated mixture, pH <2*
- Characterization
  - XRD
  - 4-point probe: Rs → conversion to thickness, assuming the initial thickness is 50 nm
  - AFM: surface roughness

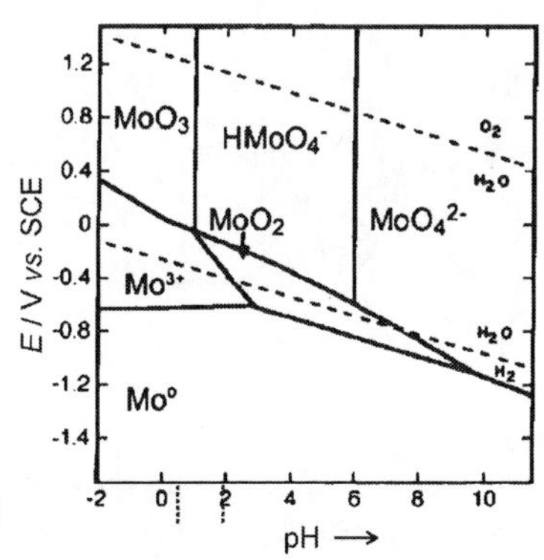

EFFECT OF TEMPERATURE ON Rs

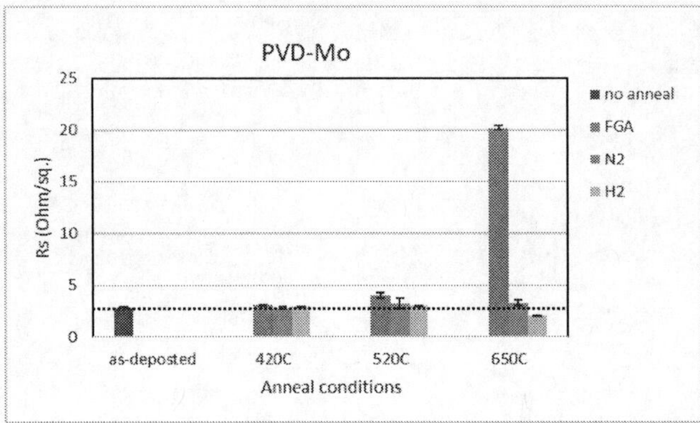

- FG anneal
  - Rs increased as a function of annealing temperature; substantial increase for 650 °C anneal
- N_2 anneal: slight increase of Rs for higher temperatures within 520-650 °C range
- H_2 anneal: significant decrease of Rs for 650 °C anneal

imec

17

GI-XRD

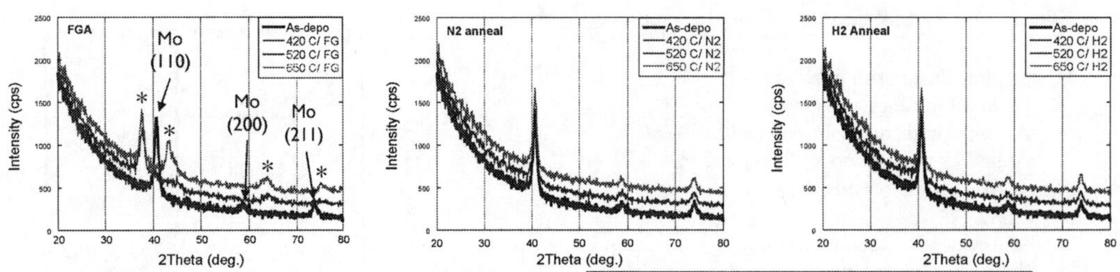

- FG anneal
  - 420 °C : no noticeable change vs. as-deposited Mo
  - 520 and 650 °C: formation of Mo nitride (*)
- N_2 and H_2 anneals: no noticeable change observed up to 650 °C

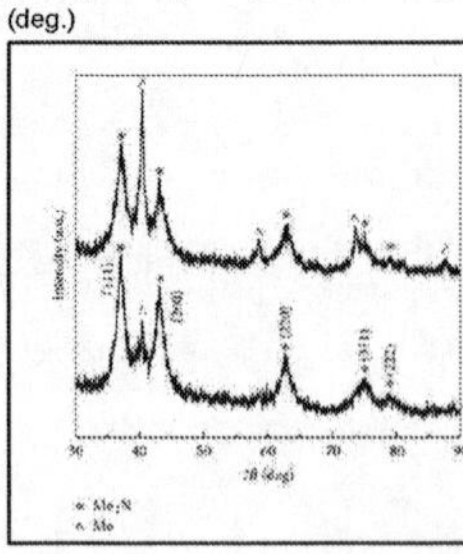

L. Shen et al., J. Nanomater. (2011) doi:10.1155/2011/781935

imec

18

EFFECT OF OXIDIZER CONCENTRATION ON Mo ETCHING

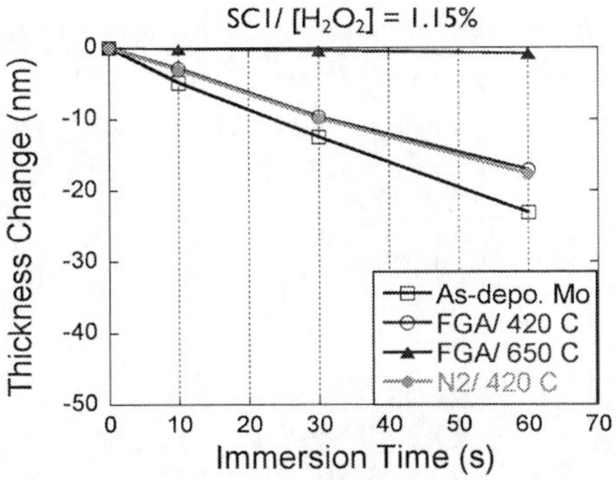

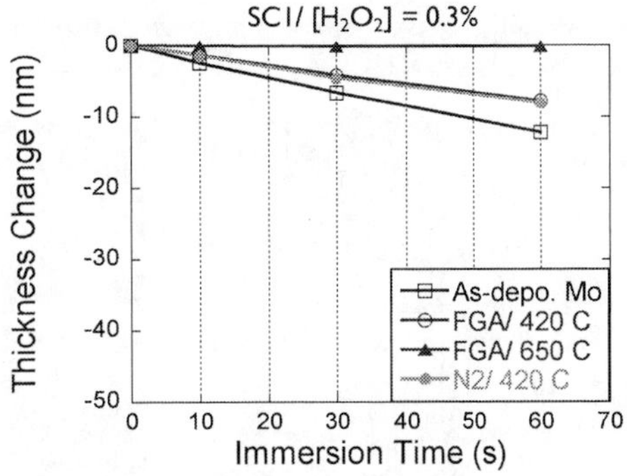

- Lower etch rate for lower concentration of the oxidizer (H_2O_2) in the mixture
- FG annealed Mo/ 650 °C: very low etching explained by the formation of Mo nitride
- Etch rate: as-depo. Mo > annealed Mo

19

FG ANNEALING vs. AS-DEPO Mo

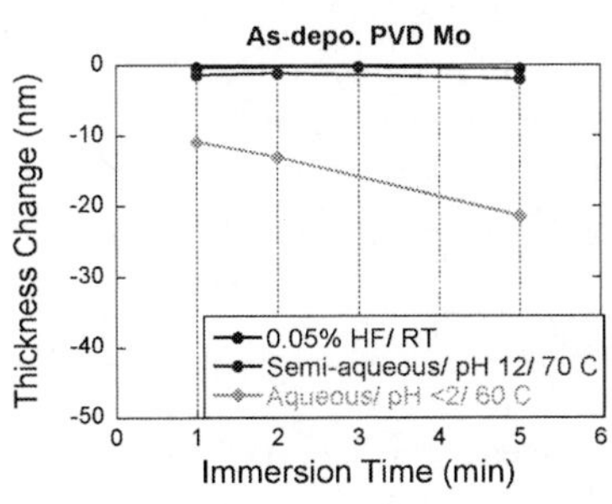

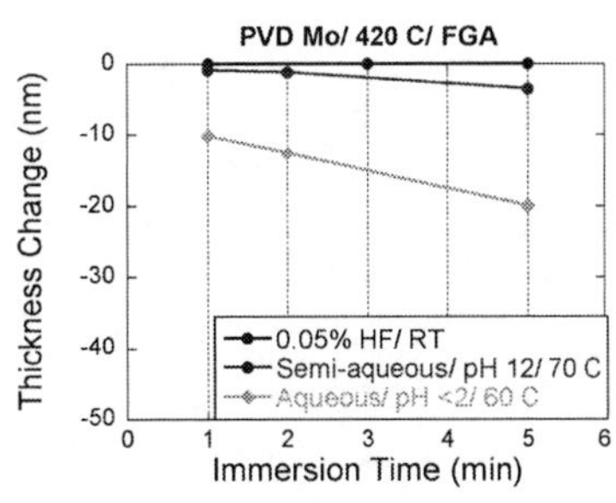

- As-depo and FGA
  - 0.05% HF and semi-aqueous mixture pH >12: no or very low etching
    - *Could be used for Mo cleaning; residues removal efficiency needs to be assessed*
  - Aqueous mixture pH <2: ~10 nm/min etch (as-depo and FGA/ 420 °C)
- Similar trend for N_2- and H_2-annealed Mo (results not shown)

20

SURFACE ROUGHNESS EVALUATION (AFM)

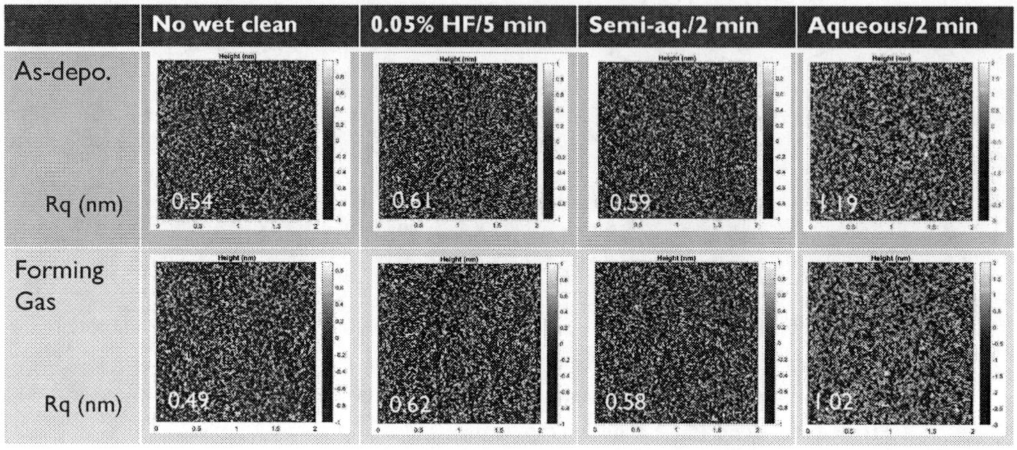

Low etching = minor increase of roughness

imec

SUMMARY

- Annealing: formation of Mo nitride if Mo was annealed at temperatures ≥ 520 °C in forming gas, explained by a lower dissociation energy of N_2 molecule with the presence of H_2 in the mixture
- SC1: substantial etch of Mo; etch rate increased with the concentration of the oxidizer (H_2O_2) in the mixture
- 0.05% HF
  - Almost no Mo etching after 5 min
  - Could be used for PERR of Mo-containing stack, depending on (a) the stack structure (dielectrics) and (b) the plasma used for patterning
- Semi-aqueous mixture pH >12
  - Very low ER for Mo, typical ER ≤0.5 nm/min
  - Could be used for residue removal of Mo-containing stack; residue removal efficient needs to be assessed

imec

imec

embracing a better life

imec PUBLIC

WETTING OF DEEP HYDROPHILIC NANOHOLES BY AQUEOUS SOLUTIONS

GUY VEREECKE[1]*, AUDREY DARCOS[2], HIDEAKI IINO[3], FRANK HOLSTEYNS[1], AND EFRAIN ALTAMIRANO SANCHEZ[1]

[1]IMEC, KAPELDREEF 75, 3001 LEUVEN, BELGIUM *guy.vereecke@imec.be

[2]INSA, 26 AVENUE DE LESPINET, 31400 TOULOUSE, FRANCE

[3]KURITA WATER INDUSTRIES LTD., 1-1, KAWADA, NOGI-MACHI, SHIMOTSUGA-GUN, TOCHIGI, 329-0105, JAPAN

PUBLIC

NANOCONFINEMENT IN SEMICONDUCTOR MANUFACTURING

Logic Fin & Nano Sheets FET

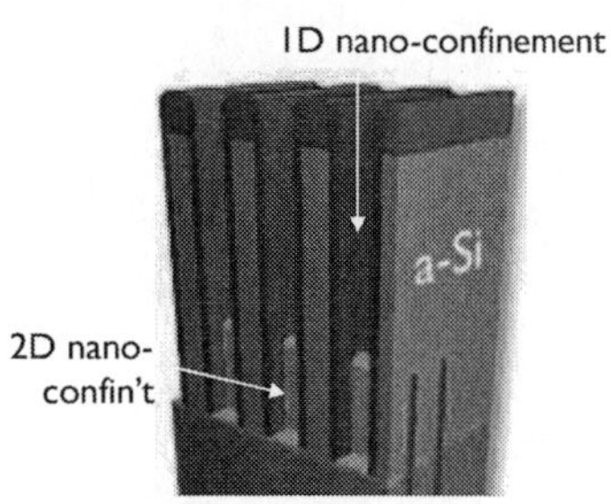

- Post-etch cleans
- STI oxide recess
- Selective semiconductor etches
- RMG back etches

Courtesy: Y. Oniki, imec

Logic BEOL Supervias

Supervia from M3 to M1 at 3 nm node

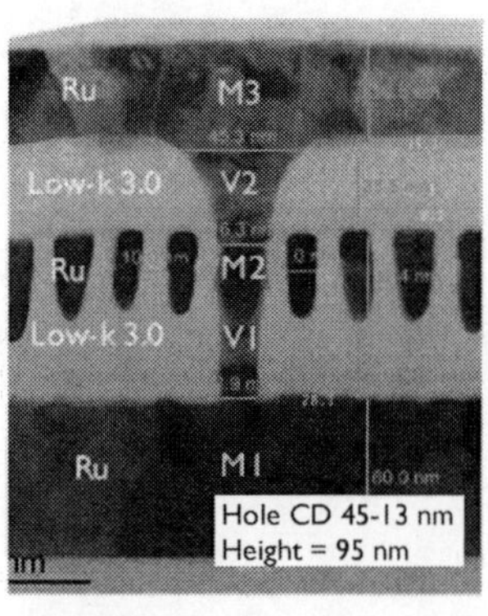

Source: V. Vega-Gonzalez, IEDM Tech. Digest (2019)

3D-NAND memory

Source: LAM Research

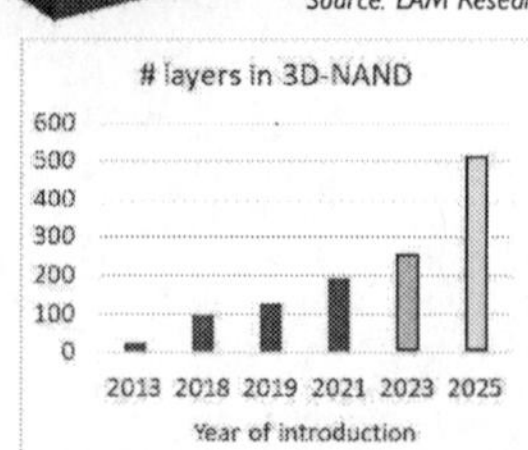

Source: S. S.-W. Wang, Semicond. Eng. (2018)

WETTING AT THE NANOSCALE
SIMULATIONS PREDICT FAST WETTING

- Fluid transport is diffusion-limited in nano-structures
- No impact of convection at wafer surface

Wetting time of a 1 µm via

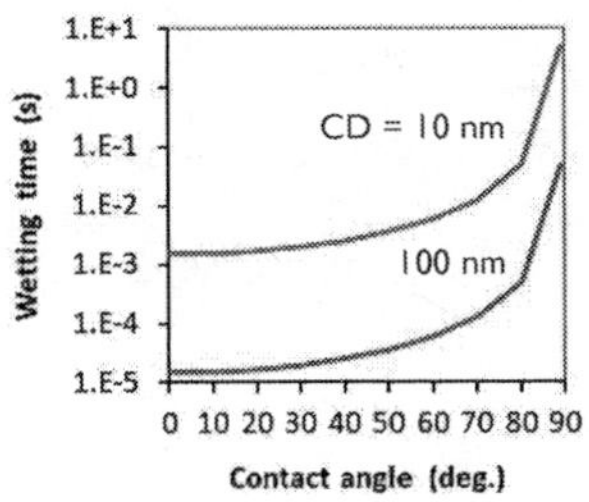

- Short wetting times from fast diffusion
- Diffusivity not expected to be an issue

Solute	D (1E-9 m^2/s)	Diffusion rate (µm/s)
H^+	9.31	136
Cu^{2+}	0.71	38
Cl^-	2.03	64
SO_4^{2-}	1.07	46

T = 25°C

- Diffusion rates calculated from the average squared net diffusion distance: $|x| = \sqrt{2Dt}$

imec Source: M.T. Fuller & D.W. Hess, JECS (2003) 3 PUBLIC

WETTING CHARACTERIZATION

- Microscope inspection – nanofluidics

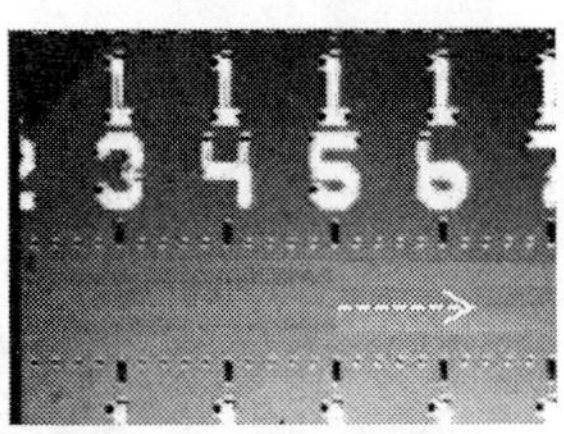

Moving meniscus in a 11 nm deep, 20 µm wide, nanochannel

→ Apparent viscosity increased by up to 4

- Not applicable to nanoelectronic structures

Sources: V.N. Phan, Langmuir (2010), K. Mawatari, Anal. Chem. (2014)

- High-frequency acoustic reflectometry

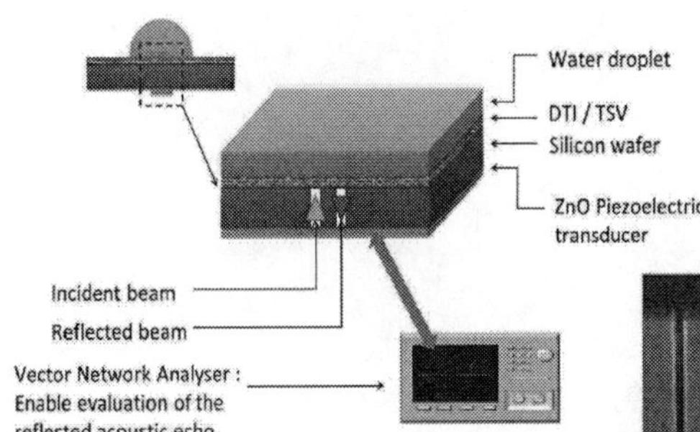

→ Detection of partial wetting in Deep Trench Isolation structures – AR ~ 20, CD = 200 nm

imec 4 Source: C. Virgilio et al., Solid State Phenom. (2016) PUBLIC

CHARACTERIZATION OF WETTING BY ATR-FTIR

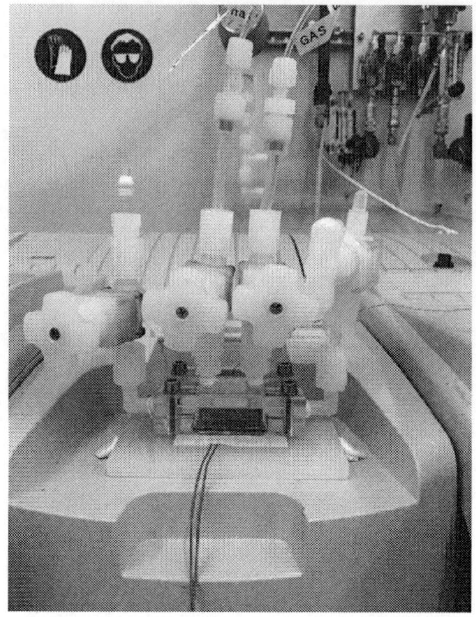

Nicolet 6700 FT-IR spectrometer
MCT detector cooled by liquid N2
Customized flow cell on ATR accessory

CHARACTERIZATION BY IN-SITU ATR-FTIR

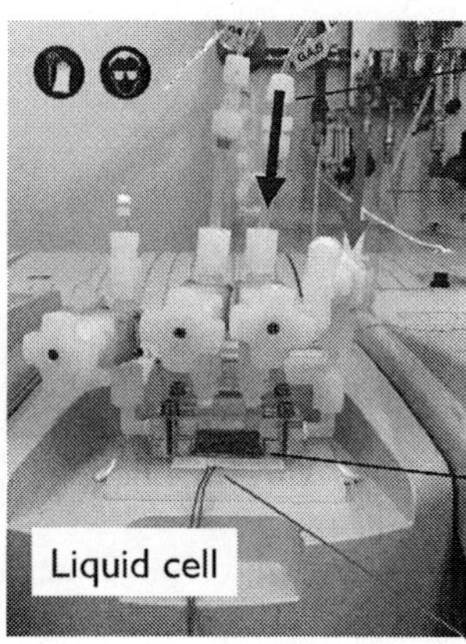

Gas inlet
- N_2 for in-situ drying
- CO_2 to characterize diffusivity & permittivity

Solution injection port

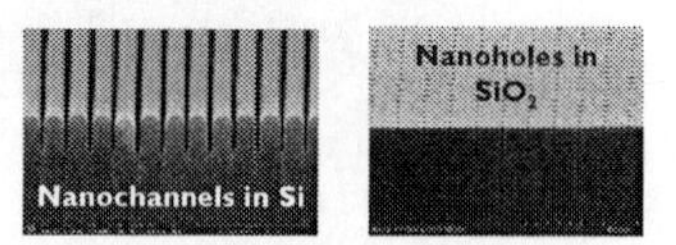

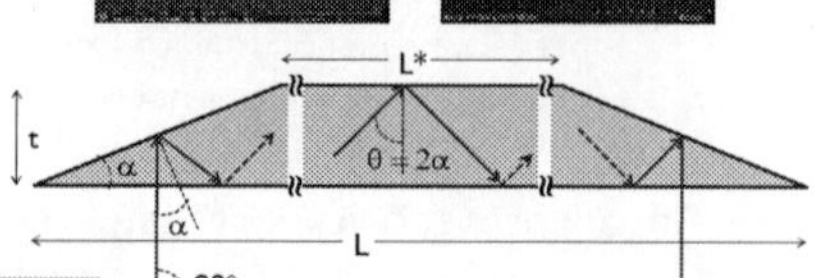

Si ATR crystal
- with blanket films
- with nano-channels/holes

Heating mattress (RT – 90°C)

- Backside heating
- ΔT BS / surface solution = 10°C at 90 °C

imec

CHARACTERIZATION OF WETTING BY IN-SITU ATR-FTIR

DETERMINATION OF THE OH STRETCHING / BENDING RATIO

Wetting of nanopillars by UPW

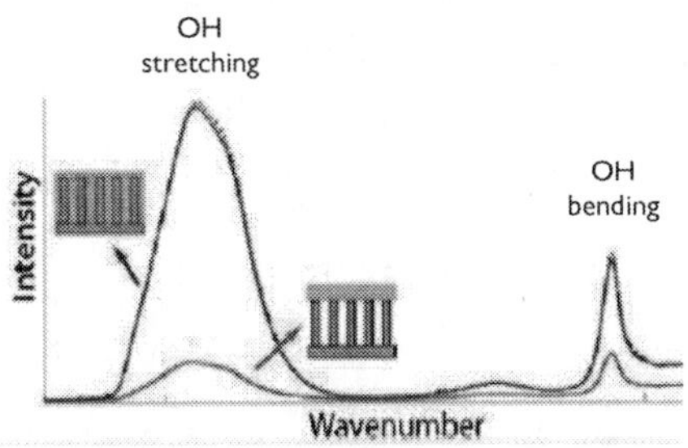

I stretching / bending ↗ upon wetting

Penetration depth of evanescent wave at Si / H_2O – θ = 60°

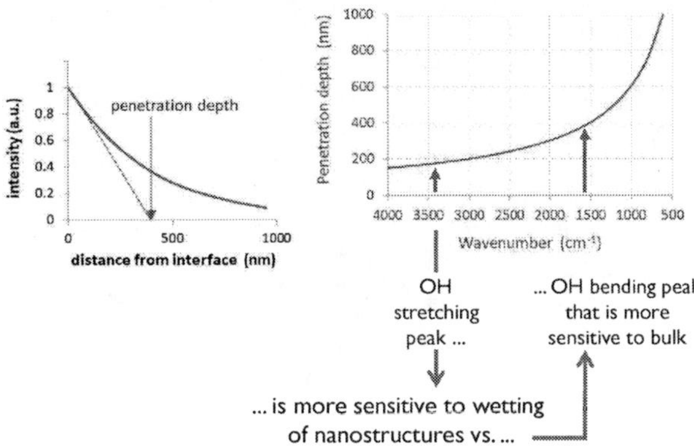

CHARACTERIZATION OF WATER STRUCTURING

- OH stretching peak

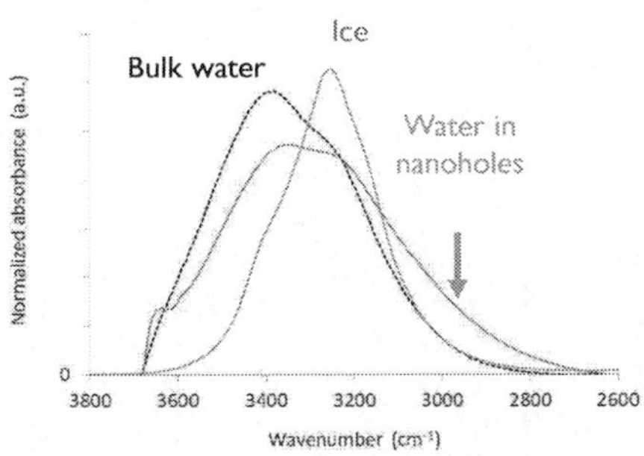

- Water in nanoholes presents a band at much lower frequency vs. ice

- Interpretation based on number of H-bonds

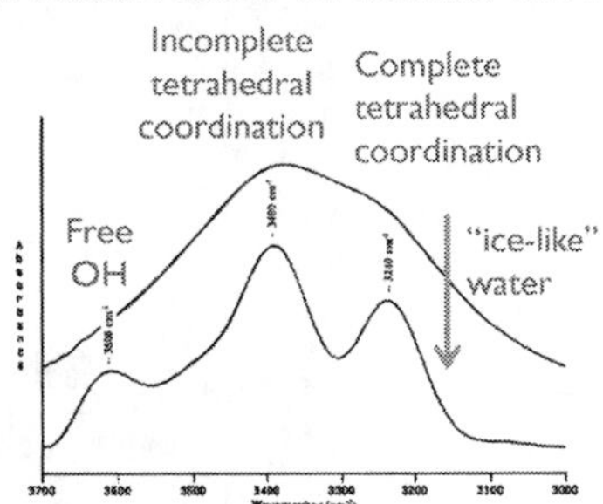

Yalamanchili et al., *Langmuir* (1996)

- But most H_2O molecules are engaged in 4 H-bonds (Chaplin, http://www1.lsbu.ac.uk/water)

→ Interpretation based on stiffness of H_2O network ?

WATER STRUCTURING BY DISSOLVED IONS

- Bulk solutions

ΔG_{HB}	Ions	Category
-1.1	I^-	Structure breaking ions
-0.9	Br^-	
-0.5	Cl^-	
-0.1	SO_4^{2-}	
0.1	Na^+	Borderline ions
0.7	Ca^{2+}	Structure making ions
1.1	PO_4^{3-}, Co^{2+}	
	Fe^{2+}	

ΔG_{HB} : average change in the number of hydrogen bonds per water molecule

Source: Y. Marcus, Chem. Rev. (2009)

- Salt solutions in 32-nm nanochannels

Difference ATR-FTIR spectra 1 M salt solution - water

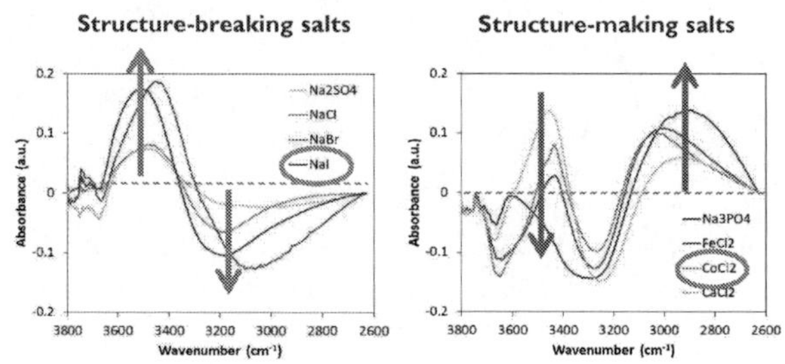

- Trends afo salts in coherence with ions in solutions
- Interpretation as $\Delta_{\text{H-bonds}}$ is doubtful over large distances

PUBLIC

WETTING OF DEEP NANOHOLES

Holes in PEALD SiO_2 on Si
Depth ~ 300 nm
CD ~ 20 nm
Volume ~7%

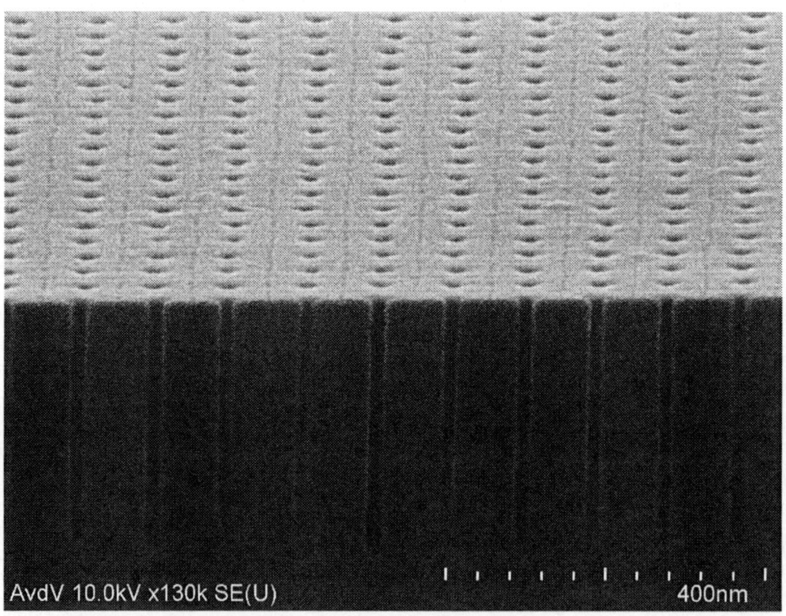

EVIDENCE FOR UNCOMPLETE WETTING

- OH stretching / bending ratio vs. T & time

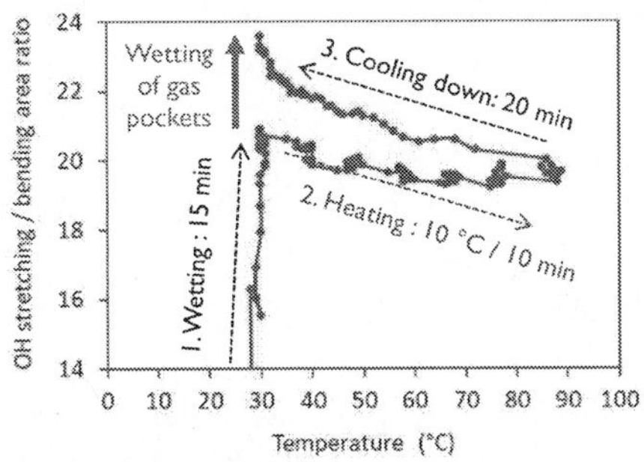

- Hysteresis is proof of uncomplete wetting
- Generation of gas pockets / nanobubbles with unexpected long lifetime[1]

- Water structuring when wetting nanoholes

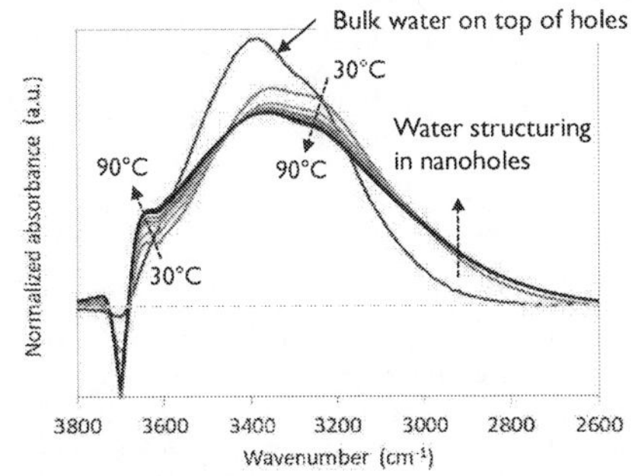

- Heating accelerates diffusion & modifies water structure, but does not suppress water structuring

CO_2 DISSOLUTION IN NANOHOLES FILLED WITH UPW

P_{CO2} = 1 atm

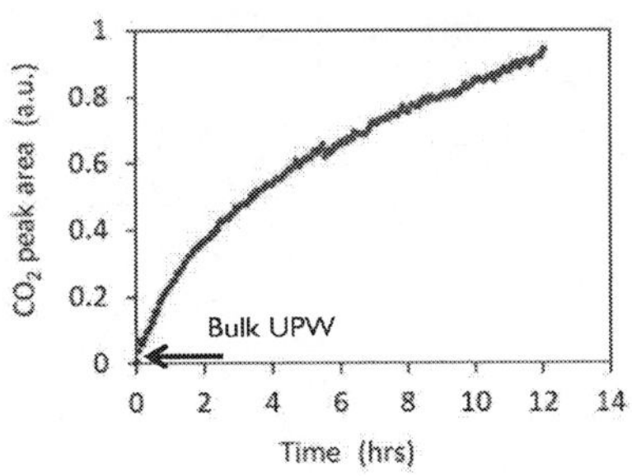

- No equilibrium after 12 hrs
  - About 10 min needed to achieve equilibrium in bulk UPW
  - Decreased diffusivity likely due to water structuring[1]

- High solubility
  - About 50 X level of bulk UPW
  - Likely due to decrease of permittivity from water structuring

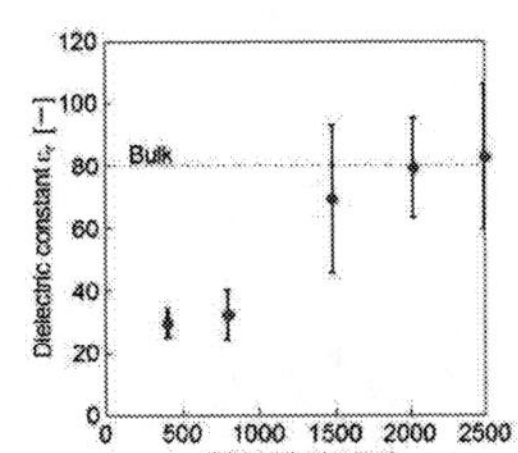

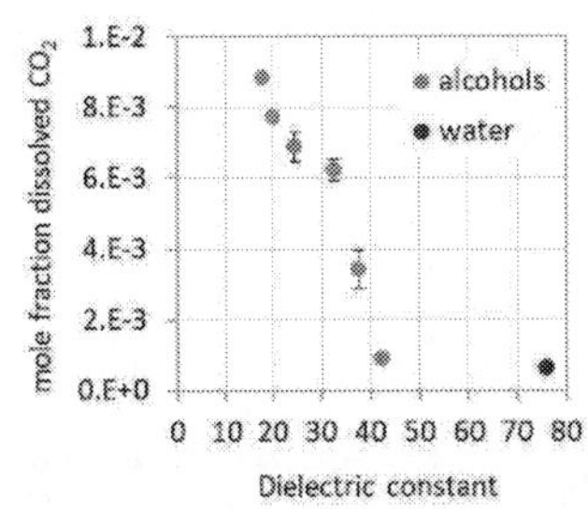

[1] T. Tsukahara et al., *J. Phys. Chem. B* (2009)

K. Morikawa et al., *Anal. Chem.* (2015)

P. Fogg, Solubility Of Gases In Liquids (1991)

WETTING KINETICS

- Characterization

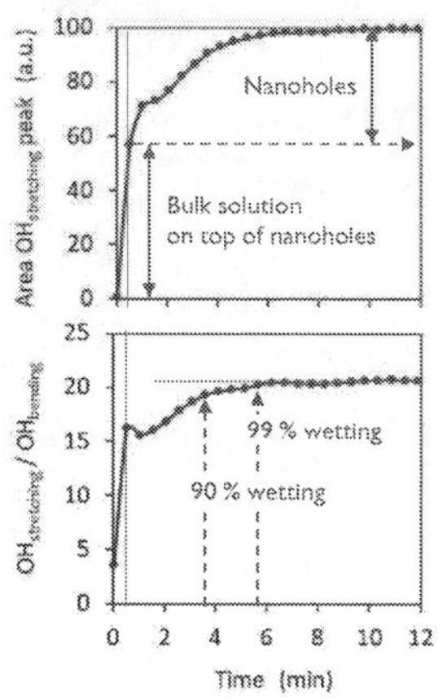

- Ion effect at RT – 30°C

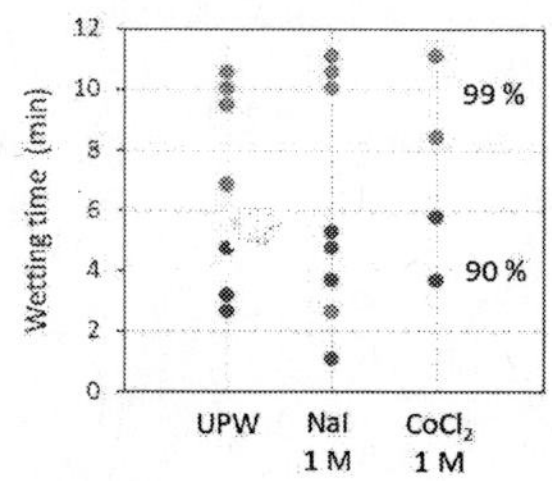

- Large variability likely from random formation of gas pockets
- Long wetting times
  - Not practical for manufacturing
  - No significant effect of salt addition

- Temperature effect

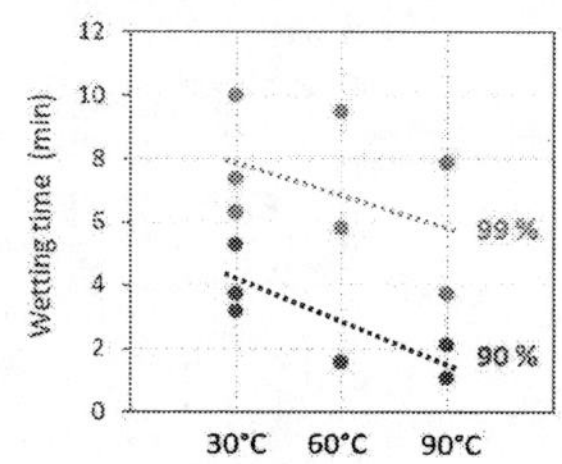

- Large variability but significant effect: faster wetting at higher T
- Heating needed for manufacturability
- Only backside heating used

EFFECT OF DISSOLVED SALTS ON WATER STRUCTURING AT RT

- OH stretching peaks

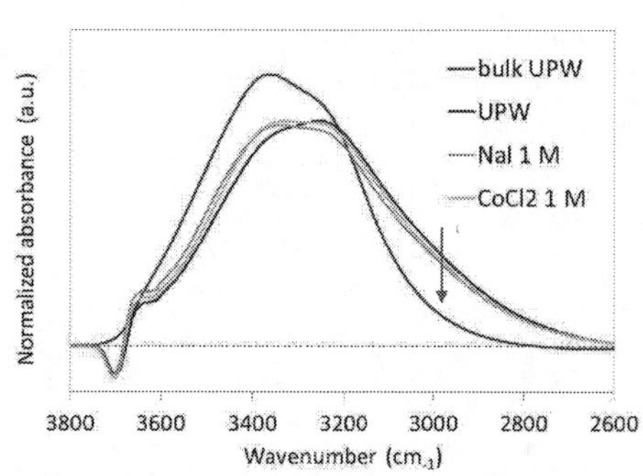

- Slight decrease of structuring by dissolved salts

- Difference spectra salt – UPW

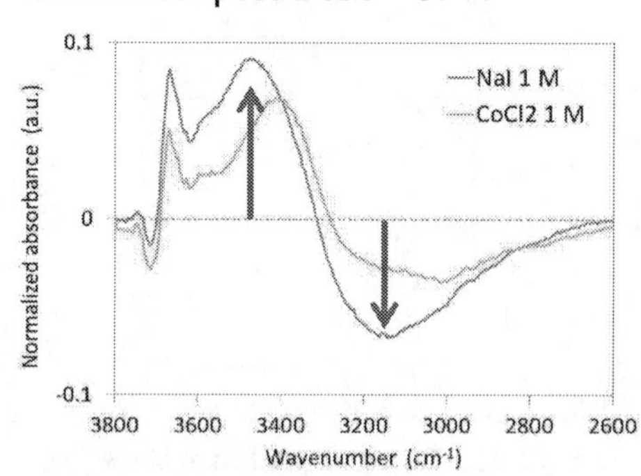

- Both salts showed structure breaking charateristics

EFFECT OF DISSOLVED SALTS ON DISSOLUTION OF CO_2

- Dissolution of CO_2 at RT & 1 atm P_{CO2}

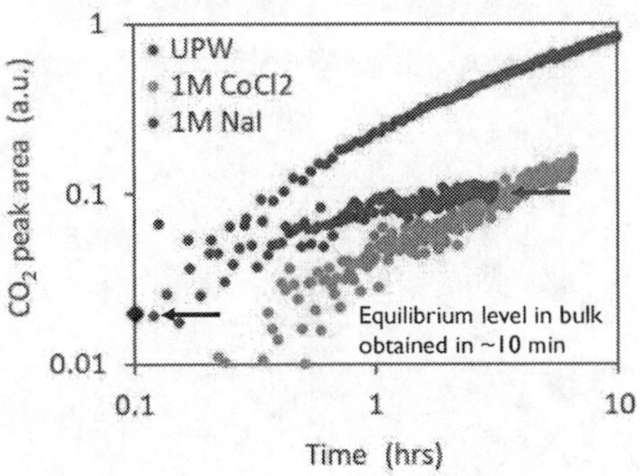

- Equilibrium achieved in ~ 2 hrs with NaI

- Lower solubility in NaI solution vs. UPW, indicating some restoration of water properties using a structure breaking salt – but slower diffusion
- Slower diffusion in $CoCl_2$ solution, with global structure making properties – but structure breaking characteristic in FTIR
- No simple relation btw structure breaking / making properties and characterisation by FTIR

imec PUBLIC

SUMMARY

- Wetting of 20 nm nanoholes in an oxide matrix was accompanied by the formation of gas pockets / nanobubbles,
  likely stabilized by slow gas diffusion from water structuring
- Slower wetting and diffusivity & lower permittivity from water structuring
  - Higher solubility of CO_2 (gases) (and lower solubility of salts) from decreased permittivity
  - Partial restoration of water permittivity by the dissolution of salts , but not of diffusivity and wetting rate
    - More salts to be tested
  - Wetting rate increased by backside heating
    - Solution heating needed for manufacturing depending on hole depth
    - But nanobubbles issue not solved

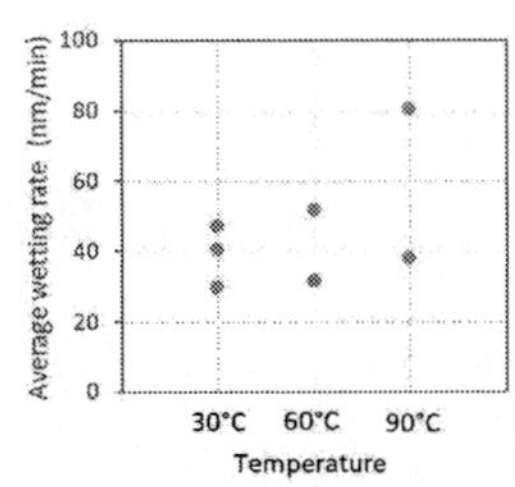

imec 16 PUBLIC

imec
embracing a better life
imec
CONFIDENTIAL

AUTHOR INDEX